MW00441053

About Island Press

Since 1984, the nonprofit organization Island Press has been stimulating, shaping, and communicating ideas that are essential for solving environmental problems worldwide. With more than 1,000 titles in print and some 30 new releases each year, we are the nation's leading publisher on environmental issues. We identify innovative thinkers and emerging trends in the environmental field. We work with world-renowned experts and authors to develop cross-disciplinary solutions to environmental challenges.

Island Press designs and executes educational campaigns, in conjunction with our authors, to communicate their critical messages in print, in person, and online using the latest technologies, innovative programs, and the media. Our goal is to reach targeted audiences—scientists, policy makers, environmental advocates, urban planners, the media, and concerned citizens—with information that can be used to create the framework for long-term ecological health and human well-being.

Island Press gratefully acknowledges major support from The Bobolink Foundation, Caldera Foundation, The Curtis and Edith Munson Foundation, The Forrest C. and Frances H. Lattner Foundation, The JPB Foundation, The Kresge Foundation, The Summit Charitable Foundation, Inc., and many other generous organizations and individuals.

The opinions expressed in this book are those of the author(s) and do not necessarily reflect the views of our supporters.

Green Growth That Works

Green Growth That Works: Natural Capital Policy and Finance Mechanisms Around the World

Edited by
Lisa Mandle, Zhiyun Ouyang,
James Salzman, and Gretchen C. Daily

Washington | Covelo | London

Copyright © 2019 Island Press

All rights reserved under International and Pan-American Copyright Conventions. No part of this book may be reproduced in any form or by any means without permission in writing from the publisher: Island Press, 2000 M Street, NW, Suite 650, Washington, DC 20036 No copyright claim is made in the works of Rick Bennett, employee of the United States federal government.

Library of Congress Number: 2018967529

All Island Press books are printed on environmentally responsible materials. ✪

Manufactured in the United States of America

10 9 8 7 6 5 4 3 2 1

Keywords: Island Press, natural capital, ecosystem services, green economy, green development, inclusive green growth, payment for ecosystem services, payment for environmental services, gross ecosystem product, environmental policy, environmental finance, conservation policy, conservation finance, sustainable development, sustainable growth, policy instrument choice, poverty alleviation, livelihood security, human development, equity, green finance, transformation

Dedicated to all the heroes reaching well beyond their own lives to help realize a bright future for people and nature on our shared planet Earth.

Contents

PART I

Introduction and Background

CHAPTER 1

The Case and Movement for Securing People and Nature

Lisa Mandle, Zhiyun Ouyang, James Salzman, Ian Bateman, Carl Folke, Anne D. Guerry, Cong Li, Jie Li, Shuzhuo Li, Jianguo Liu, Stephen Polasky, Mary Ruckelshaus, Bhaskar Vira, Alvaro Umaña Quesada, Weihua Xu, Hua Zheng, and Gretchen C. Daily

Growth in human population and economic activity has dramatically transformed our planet since the Industrial Revolution. While driving significant improvements in human well-being, these forms of growth also deeply eroded the natural capital embodied in Earth's lands, waters, and biodiversity. Rapid economic development has lifted hundreds of millions of people out of poverty and raised the standard of living and life expectancies of many more, but the costs of this success cast a long shadow over future well-being.

Massive degradation and loss of forests, wetlands, coral reefs, grasslands, and other ecosystems—along with their waters and species—are creating grave risks. Severe inland and coastal flooding, sand and dust storms, extreme climate events, and unhealthy air and water threaten the security of food, water, climate, energy, health, and livelihoods. Taken together, these harms constitute a sobering counterbalance to economic growth.

Yet the world is waking up. Policymakers, development organizations, businesses, and private investors are joining civil society organizations and scientists in forging a new development model. The quest is ambitious: to improve the human condition while securing nature's life-support systems—in other words, to create pathways toward inclusive green growth.

Over two decades of innovation in research and implementation demonstrate the feasibility of this ambitious yet urgent and vital goal. The challenge now is to move from ideas to action on a broader scale (Guerry et al. 2015). Doing so requires a clear focus on values and mechanisms.

Moving to inclusive green growth necessitates a rigorous understanding of the significance and values of natural capital for human well-being—today and for the future. Whether in the public or private sector, decision makers will need to be persuaded of the benefits from major investments in nature and nature-based solutions compared to their costs. This understanding has been advanced rapidly through scientific research (going back to Jansson et al. 1994 and Daily 1997), as well as via international efforts such as the Millennium Ecosystem Assessment (2005) and the Intergovernmental Science-Policy Platform on Biodiversity and Ecosystem Services (IPBES).

Going from ideas to action also requires creating innovative, effective policy and finance mechanisms that drive regeneration and promotion of natural capital and the provision of ecosystem service benefits (Daily and Ellison 2002). These approaches typically have deeply intertwined, dual goals of securing the well-being of both nature and people.

In this book, we put the emphasis on the natural capital dimensions underpinning inclusive green growth, highlighting the recent, rapid innovation in demonstrating the economic and national security—as well as the inclusive and ethical—case for securing nature. And we emphasize the experience in diverse countries worldwide in implementing innovative policy and finance mechanisms. Our focus on natural capital complements other efforts around inclusive growth (de Mello and Dutz 2012), inclusive development (Teichman 2016), and green economies (Jacobs 1993).

Recognition of the value of natural capital is not new: one can find examples of wise conservation practices from societies throughout history. Yet, today, many human activities not only fail to consider natural capital but actively degrade its value. Because services from natural capital are generally public goods and not reflected in market prices, their loss is often unrecognized—or at least unchecked—until the consequences become too large to ignore. A famous case is China, where massive deforestation in the upper reaches of the Yangtze River led to devastating flooding in 1998. In 1999, China began to launch what is still the largest payment for ecosystem services program in the world—the Sloping Land Conversion Program—with dual goals of flood mitigation and poverty alleviation (Liu et al 2018).

Society's most vulnerable members often have the greatest immediate dependence on nature, and the lowest ability to cope with or substitute for loss of ecosystem service benefits (Gadgil and Guha 1992; MEA 2005). In the face of natural disasters and extreme climate events, poor and marginalized individuals and communities are often most likely to end up in harm's way (Hamann et al. 2018). Thus the deterioration of natural capital often hits the most vulnerable people the hardest, making considerations of inclusivity and equity imperative in the context of ecosystem services (Vira 2015). Ultimately, however, all people depend on nature for their well-being. Although wealth provides a buffer, no one is secure from the escalating risks of continuing on our current path of depleting natural capital.

As the consequences of natural capital losses have grown more pronounced, their impact has sparked innovations over the past two decades by visionary leaders around the world. In 2000, the term *ecosystem services* was scarcely known. There are now over 500 payment for ecosystem services (PES) programs around the globe with annual payments exceeding US$36 billion (Salzman et al. 2018). Such programs have become a central component of China's nationwide human development, environmental protection, and national security strategy (Ouyang et al. 2016; Bryan et al. 2018). These innovations initially took the form of custom, one-off mechanisms. Over time, they are developing into more complete, tested tools and strategies on which policymakers can draw with confidence.

One question in these efforts has remained central: how can economic and other incentives be designed to make investments in natural capital attractive and commonplace? The legal regimes in place to date have been primarily regulatory, relying on prescriptive statutes to direct behavior. These are important, but have clearly proven inadequate. Focusing on the incentive structure for landholders and other ecosystem service providers—the stewards of natural capital stocks of forest, wetlands, grazing lands, and coral reefs—provides a complementary strategy that is just now starting to reach its potential.

Accurate valuation of natural capital and effective local mechanisms to protect and restore it are necessary, but not enough. The overarching challenge lies in achieving action at a magnitude and extent capable of effecting global change. This requires scaling and adapting these approaches so that they become mainstream. Fortunately, recognition of the need for action is growing, and momentum behind transformation is building. Interest from governments, private institutions, and society at large is evident globally in a number of international initiatives.

The 2030 Agenda for Sustainable Development, adopted in 2015 by governments worldwide, sets out a universal agenda for achieving the triple bottom line of sustainable development—economic, social, and environmental (Elkington 1998). The year 2020 will see an update of the Convention on Biological Diversity's strategic plan, providing what is hoped to serve as a "New Deal for Nature": a framework for the international community to "ensure that the solutions and benefits nature provides are integrated in systemic, inclusive, and transformative actions to benefit human well-being, the economy, and the planet." The Natural Capital Coalition has, since 2014, brought together nearly 250 businesses and other organizations from around the world to advance the vision for a world where businesses conserve and enhance natural capital. The Natural Capital Project, a partnership among scientific research institutions and conservation organizations, works to mainstream the value of nature in decisions made by governments, businesses, multilateral development banks, and others. It has worked directly in over 60 countries. The Natural Capital Project's data and science platform, anchored by the InVEST software suite, has been used in over 180 countries to map, quantify, and value ecosystem services in support of a diversity of policy and finance mechanisms.

Reflecting the growing demand to collect and describe important policy and finance solutions to securing natural capital, this book has its origins in a request from the leadership in China's central government. In 2017, the Department of Development Planning of the National Development and Reform Commission asked for a report on international experience in green finance and development. Chinese President Xi Jinping has declared that it is "China's dream" to become "the ecological civilization of the 21st century." This means pioneering, testing, and implementing a comprehensive, transformational system of policy and finance mechanisms that deliver improvements in human well-being—especially in poor and vulnerable communities—and that conserve and restore ecosystems and their ability to provide goods and services now and into the future.

In drafting the report, we found many examples of important policy and finance innovations for green and inclusive development. But these were scattered across the academic literature, white papers, and reports, or locked in the minds of the leaders who shepherded them from idea to implementation. In synthesizing this information for China's leaders, we realized it would usefully serve a wider, global audience.

This book has been designed to provide a practical guide to how policies and finance mechanisms have been implemented in the real world, across a diversity of contexts, in order to secure and enhance natural capital and ecosystem service benefits on the pathway toward inclusive green growth. Through a series of case studies, we address key questions such as the following: How can economic and other incentives be channeled to make conservation and restoration attractive and mainstream? Which approaches have been successful, under what conditions? What roles can governments, businesses, landowners, nongovernmental organizations, and other actors play, and how can they complement each other? What are the opportunities to scale and improve these successful models?

In the remainder of this chapter, we introduce several key concepts (box 1.1) that run throughout the rest of the book, provide an overview of the book's format and organization, and make suggestions for steps forward.

How This Book Is Organized

This book surveys the range of policy and finance mechanisms that channel economic resources and other benefits toward securing and enhancing natural capital. These mechanisms typically also aim to increase equity and well-being, both through poverty alleviation and in access to ecosystem goods and services. Illustrative examples have been contributed by a range of experts who come from the natural and social sciences, government, private companies, financial institutions, and civil society organizations. Authors have been involved in all aspects of policy and finance, from design and implementation to evaluation and scaling up these mechanisms—and the examples come from both developing and developed countries worldwide.

In part 1, we provide an overview of the state of practice and implementation of initiatives to integrate natural capital into societal decisions, and of pathways to scaling mechanisms toward global change.

Part 2 presents six chapters, each focusing on a specific policy or finance mechanism for conserving or enhancing natural capital. These include government subsidies, regulatory-driven mitigation, voluntary conservation, water funds, market-based transactions, and bilateral and multilateral payments (figure 1.1). To ground the discussion in practice and demonstrate the wide range of approaches around the globe, every chapter provides a set of detailed case studies. To make comparisons straightforward, the case studies follow a common format with the following sections, when applicable:

Key Concepts

Inclusive green growth is "efficient in its use of natural resources, clean in that it minimizes pollution and environmental impacts, and resilient in that it accounts for natural hazards and the role of environmental management and natural capital in preventing physical disasters. Importantly, green growth is not inherently inclusive. Its outcome will likely be good for the poor, but specific policies are needed to ensure that the poor are not excluded from benefits, and are not harmed in the transition. The welfare impacts of green policies will be greater if efforts are made to make the policies inclusive" (World Bank 2012, p. 30). Such an approach is sometimes labelled *inclusive green development* or *green economy* (UNEP 2011). The common principles are reducing poverty and improving access to health, education, and infrastructure services, while investing in the natural assets on which livelihoods and economies depend.

Natural capital refers to stocks of nonliving and living elements of ecosystems that provide benefit streams to people. Prominent nonliving natural capital assets include ores and minerals. We are focused here primarily on living stocks of natural capital, including ecosystem assets such as fertile soils, forests, coastal marshes, farmland, and the diversity of life living in ecosystems.

Ecosystem goods and services flow from stocks of natural capital, often coproduced in combination with flows from other capital, including human labor, ingenuity, and manufactured goods. These benefits—also referred to as nature's contributions to people (Díaz et al. 2018)—are vital to human well-being and, indeed, our very existence. They include everything from food, fuel, and fiber; to the provision of clean water; to mitigation of extreme events such as coastal storms and inland flooding. Ecosystem services also include a diversity of nonmaterial benefits, such as psychological and physical health, cultural identity, sense of place, and nature-based tourism and recreation.

Interactions and trade-offs are inherent in the dynamics of ecosystems, where everything is interconnected. It is therefore key to analyze trade-offs in the production of ecosystem goods and services (Howe et al. 2014). For example, timber can be a valuable source of income, but timber extraction can reduce an area's recreational value and water quality regulation services, leading to increased sediments in downstream water bodies.

Resilience is the capacity of a system to retain essential structures, processes, and feedbacks in the face of shocks or disturbances and continue to develop (Walker et al. 2004). This capacity includes linked social and ecological dimensions, such as the regenerative ability of ecosystems and their ability to deliver ecosystem services in the face of change, as well as the system's capacity for self-organization, learning, and adaptation (Folke et al. 2002).

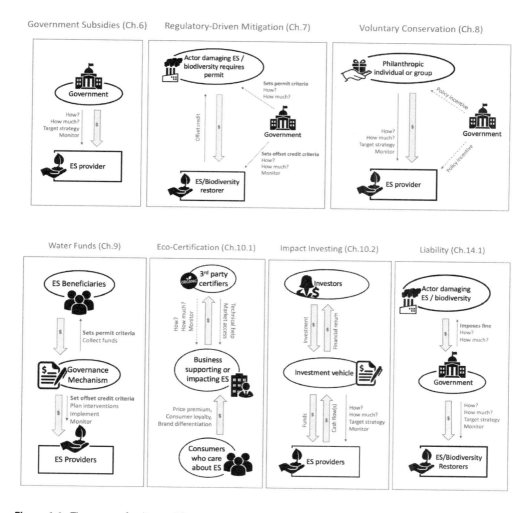

Figure 1.1. The range of policy and finance mechanisms for securing and enhancing natural capital in support of inclusive green growth.

The Problem explains the challenge that the policy seeks to address. It may be contamination of drinking water sources, loss of biodiversity, or rural poverty. This initial framing is important because choosing the most appropriate policy instrument hinges on the nature of the specific problem.

The Ecosystem Service(s) describes the biophysical basis for how the benefit will be provided. Not all natural capital is equally effective in service provision. Focusing on upstream forest conservation may be more effective to ensure flood protection and water quality, for example, than downstream grasslands.

Ecosystem Service Beneficiaries sets out which parties stand to either benefit from service provision or risk experiencing the loss of service. Payment schemes are driven by demand: by the perceived scarcity of ecosystem services. The scarcity may concern water quality, flood protection, climate stability, or loss of biodiversity. Because many services are public goods, demand may need to be amplified through regulation. This prevents free-riding and overcomes the collective action costs of organizing diffuse beneficiaries.

Ecosystem Service Suppliers sets out who provides the service. In most cases, this is landowners. To change their behavior, the incentives must be competitive with the opportunity costs, to make conservation as attractive as, for example, the values of timber or palm oil plantations.

Terms of the Exchange: Quid Pro Quo refers to the conditions of the exchange. If payments are involved, what are the respective obligations of the parties? For example, is the landowner paid for the input (changing the land management) or the output (actual provision of the desired service)?

Mechanism for Transfer of Value describes the operation of the exchange. For subsidy programs, how is eligibility determined and how are payments distributed? Institutions are particularly important in this context in order to efficiently bring together services suppliers and beneficiaries.

Monitoring and Verification concern how those providing value or regulating service provision can ensure that the appropriate land management practices are, in fact, undertaken. This is a critically important aspect of any policy mechanism. Inadequate monitoring increases the possibility of cheating and inefficient provision of services.

Effectiveness addresses whether the program actually makes a difference. Payment and subsidy schemes may operate smoothly but not change the actual flow of services or state of natural capital. One can measure effectiveness in terms of service provision (a biophysical measure); efficiency (an economic measure); or improvement of social welfare (such as poverty reduction, gender equity, or securing property rights). For the vast majority of natural capital programs, we simply do not know because most policies are never assessed for effectiveness.

Key Lessons Learned provides a distillation of important conclusions for those interested in applying a similar approach to new contexts.

Key References point to details about the mechanisms and case studies.

In part 3, each of the six chapters focuses on policy mechanisms or strategies that have been established within specific countries or regions to conserve or enhance natural capital. The first two chapters, featuring China and Costa Rica, describe the most comprehensive, systemwide approaches to transform policy and finance to inclusive green growth that we know of worldwide, initiated in the late 1990s and operating from national to local levels. The subsequent three chapters describe inspiring innovation and frontiers in a suite of other countries. The final chapter focuses on innovation and potential now beginning to blossom in cities.

Finally, interspersed throughout the book, you will find boxes highlighting additional innovative approaches. The spread of the strategies and mechanisms covered created a not-unhappy challenge for us: we were faced with more examples than could fit as full-length case studies. We have therefore included some of the newest cases as separate boxes within chapters, and we encourage you to follow up with the key references listed to learn more, and to follow their continued advancement.

How to Read This Book

It is not necessary to read this book from front to back, cover to cover. We suggest diving in wherever best fits your interests.

Throughout the book, we highlight the varying roles played by governments, the private sector, and civil society and—critically—their often-complementary interactions.

- For examples showcasing a particularly strong role for the private sector, see chapter 10 (Market-Based Mechanisms), chapter 7 (Regulatory Mechanisms), as well as the Mongolia case in chapter 2.

- For examples of public-private partnerships, see the Amazon Region Protected Areas Program in chapter 8, and water funds in chapter 9. Government can serve as the primary actor behind the mechanism (chapter 6, Government Payments, and chapter 11, Bilateral and Multilateral Mechanisms) or can catalyze opportunities for landowners, companies, and civil society organizations through regulation (see chapter 7, Regulatory Mechanisms, for a good starting point).

The case studies span a great diversity of geographies and ecosystems. To find examples from a specific place, see figure 1.2. Most mechanisms have the flexibility to be adapted to a range of contexts, so we encourage you to think of the potential beyond the specific ecosystem services or ecosystem types of any particular case study.

- For examples from marine and coastal systems, see chapter 14, with a focus on the coastal United States, and chapter 16, focused on Caribbean countries.
- For forest-based ecosystem services, see chapter 11 (Bilateral and Multilateral Mechanisms) and chapter 13 (Costa Rica).
- Examples of mechanisms for freshwater-based services are too numerous to list comprehensively, but see water funds (chapter 9), the Water Sharing Investment Partnership in Australia (chapter 10), and stormwater management in Melbourne, Australia (chapter 17) and Washington, DC, USA (chapters 7 and 10) for several cases.

The value of ecosystem service benefits can be usefully represented for decision making in a variety of ways, from biophysical metrics, to the number of people benefited, to monetary values. When monetary valuation does play a part, it is often to account for the value of specific benefits, accompanied by an understanding of who receives and who provides those benefits, and the implementation costs relative to alternative options.

- For examples of the roles monetary valuation can play, see the case of New York City's water supply (chapter 6), the Upper Tana Water Fund (chapter 10), integrated coastal zone management in Belize (chapter 16), and Working for Water in South Africa (chapter 6).

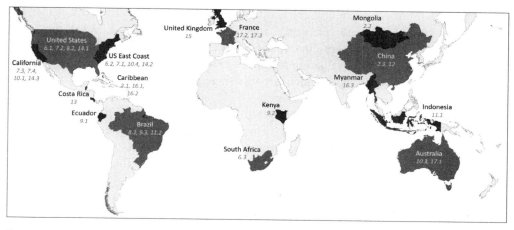

Figure 1.2. The case studies included in this book span the world's continents and represent a diversity of environmental, economic, and political contexts.

Many of these policies and mechanisms are driven by the vision of achieving inclusive green growth. To date, however, the "inclusive" part of equation—the goal of shared prosperity—has on the whole received less explicit consideration than the "green" and "growth" or "development" aspects.

- For some examples of how equity, inclusivity, and poverty alleviation goals have been integrated into these mechanisms, South Africa's Working for Water (chapter 6), Washington, DC's stormwater retention credit system (chapter 7), and multiple programs in China (chapter 12) provide a good starting place.

Looking to the Future

We hope that you will be inspired to adapt these mechanisms to new places and contexts, contributing to the scaling that is so critically needed to achieve inclusive green growth. Scaling also provides opportunities for—and, indeed, necessitates—further refinement and improvement of today's approaches. Some arenas that stand out as calling for further attention include these:

Evaluating effectiveness: Relatively few of the case studies here have been subject to comprehensive, rigorous evaluations of their effectiveness from environmental, economic, and social dimensions. This is likely in part because many mechanisms were not designed for evaluation; and in part because evaluation can be difficult in terms of the data and resources required. Evaluating outcomes may be challenging but is critical to understanding in which circumstances differing mechanisms can succeed, along which dimensions.

Increasing inclusivity: Integrating natural capital into policy and finance mechanisms has typically involved more attention to environmental and economic outcomes, with less consideration given to the issues of equity, environmental justice, and poverty alleviation. The next frontier is to improve explicit integration of these dimensions both in evaluation of policy and finance mechanisms, and in the design and implementation of new policies and programs going forward.

Diverse methods of valuation: Not all successful policy and finance mechanisms rely on an explicit calculation of the monetary value of an ecosystem service. Outcomes can also be considered in a variety of non-monetary terms such as biophysical metrics, number of people benefiting, or the equity implications. Monetary valuation may be helpful in providing a lower-bound value to include in cost-benefit analysis as for New York City's water supply (chapter 6), where watershed protection was estimated to be cheaper than the alternative of building and operating a water treatment plant. Decision making need not hinge on trying to estimate a monetary value (with questionable meaning) for the scenic beauty of these landscapes or the inspiration they provide, for example. Improved representation of nature's contributions to human well-being in nonmonetary terms, such as in terms of mental health and physical health, is an important frontier, especially for promoting inclusive benefits.

Increasing private sector engagement: Increasing contributions from the private sector toward securing and enhancing natural capital is vital to transforming to inclusive green growth. The scales—economic and geographic—at which the largest companies operate exceed those of most governments and nearly all nongovernmental organizations. Although private sector involvement is increasing, there is need for much more rapid, widespread engagement.

Science and technology frontiers: The scientific community—across both the natural and social sciences—has made tremendous progress over the past fifteen years in demonstrating the values of natural capital stocks and ecosystem service flows. But large opportunities remain to reveal the intimate connections between people and nature. Advances in sharing the growing body of global data, from remote sensing and other sources, could greatly inform policy design and implementation, for example by lowering transaction costs and increasing efficiency through better targeting of incentives and interventions.

Key References

Bryan, Brett A., Lei Gao, Yanqiong Ye, Xiufeng Sun, Jeffery D. Connor, Neville D. Crossman, Mark Stafford-Smith et al. 2018. "China's response to a national land-system sustainability emergency." *Nature* 559, no. 7713:193.

Daily, Gretchen C. 1997. *Nature's Services*. Island Press, Washington, DC.

de Mello, Luiz, and Mark A. Dutz. 2012. *Promoting Inclusive Growth: Challenges and Policies.* Paris: OECD and the World Bank.

Díaz, Sandra, Unai Pascual, Marie Stenseke, Berta Martín-López, Robert T. Watson, Zsolt Molnár, Rosemary Hill et al. 2018. "Assessing nature's contributions to people." *Science* 359, no. 6373:270–72.

Elkington, John. 1998. "Accounting for the triple bottom line." *Measuring Business Excellence* 2, no. 3:18–22.

Folke, Carl, Steve Carpenter, Thomas Elmqvist, Lance Gunderson, Crawford S. Holling, and Brian Walker. 1992. "Resilience and sustainable development: Building adaptive capacity in a world of transformations." *AMBIO* 31, no. 5:437–40.

Gadgil, Madhav, and Ramachandra Guha. 1992. *This Fissured Land*. Oxford: Oxford University Press.

Guerry, Anne D., Stephen Polasky, Jane Lubchenco, Rebecca Chaplin-Kramer, Gretchen C. Daily, Robert Griffin, Mary Ruckelshaus et al. 2015. "Natural capital and ecosystem services informing decisions: From promise to practice." *Proceedings of the National Academy of Sciences* 112, no. 24:7348–55.

Hamann, Maike, Kevin Berry, Tomas Chaigneau, Tracie Curry, Robert Heilmayr, Patrik JG Henriksson, Jonas Hentati-Sundberg et al. "Inequality and the biosphere." *Annual Review of Environment and Resources* 43:61–83.

Howe, Caroline, Helen Suich, Bhaskar Vira, and Georgina M. Mace. 2014. "Creating win-wins from trade-offs? Ecosystem services for human well-being: A meta-analysis of ecosystem service trade-offs and synergies in the real world." *Global Environmental Change* 28: 263–75.

Jacobs, Michael. 1993. *The Green Economy: Environment, Sustainable Development and the Politics of the Future*. Vancouver: University of British Columbia Press.

Jansson, AnnMari, Monica Hammer, Carl Folke, and Robert Costanza, eds. 1994. *Investing in Natural Capital*. Washington, DC: Island Press.

Kareiva, Peter, Heather Tallis, Taylor H. Ricketts, Gretchen C. Daily, and Stephen Polasky, eds. 2011. *Natural Capital: Theory and Practice of Mapping Ecosystem Services*. Oxford: Oxford University Press.

Lambin, Eric F., and Patrick Meyfroidt. 2011. "Global land use change, economic globalization, and the looming land scarcity." *Proceedings of the National Academy of Sciences* 108, no. 9:3465–72.

Liu, Jianguo. 2014. "Forest sustainability in China and implications for a telecoupled world." *Asia & the Pacific Policy Studies* 1, no. 1:230–50.

Liu, Jianguo, Andrés Viña, Wu Yang, Shuxin Li, Weihua Xu, and Hua Zheng. 2018. "China's environment on a metacoupled planet." *Annual Review of Environment and Resources*, no. 43:1–34.

MEA. 2005. *Millennium Ecosystem Assessment.* Washington, DC: Island Press.

Ouyang, Zhiyun, Hua Zheng, Yi Xiao, Stephen Polasky, Jianguo Liu, Weihua Xu, Qiao Wang et al. 2016. "Improvements in ecosystem services from investments in natural capital." *Science* 352, no. 6292:1455–59.

Salzman, James, Genevieve Bennett, Nathaniel Carroll, Allie Goldstein, and Michael Jenkins. 2018. "The global status and trends of Payments for Ecosystem Services." *Nature Sustainability* 1, no. 3:136.

Teichman, J. A. 2016. *The Politics of Inclusive Development.* New York: Palgrave Macmillan.

UNEP (United Nations Environmental Programme). 2011. Towards a Green Economy. United Nations Environment Programme.

Vira, Bhaskar. 2015. "Taking natural limits seriously: Implications for development studies and the environment." *Development and Change* 46, no. 4:762–76.

Walker, Brian, Crawford S. Holling, Stephen R. Carpenter, and Ann Kinzig. 2004. "Resilience, adaptability and transformability in social–ecological systems." *Ecology and Society* 9, no. 2.

World Bank. 2012. *Inclusive Green Growth: The Pathway to Sustainable Development.* Washington, DC.

CHAPTER 2

Scaling Pathways for Inclusive Green Growth

**Mary Ruckelshaus, Gretchen C. Daily, Stuart Anstee, Katie Arkema,
Onon Bayasgalan, Carter Brandon, Becky Chaplin-Kramer, Helen Crowley,
Marcus Feldman, Annette Killmer, Cong Li, Jie Li, Shuzhuo Li,
Michele Lemay, Jianguo Liu, Carl Obst, Zhiyun Ouyang, Stephen Polasky,
Enkhtuvshin Shiilegdamba, Samdanjigmed Tulganyam, Ray Victurine,
Greg Watson, Weihua Xu, Hua Zheng**

*The world is at a critical moment, with an urgent need to protect Earth's life-support
systems and clear avenues for how to do that. The opportunity in front of leaders from
government, multilaterals, business, research institutions, and communities is to amplify
promising demonstrations of how improved ecosystem conditions can in turn improve
human well-being. Innovation is underway in many regions, especially in China and
Latin America. Proven pathways to integrating the diverse values of natural capital
into policies and investments exist, ranging from informing infrastructure development
strategies of major development banks, governments, and private investors, in
transportation and other key sectors; to working with diverse stakeholder communities
in strategic use of land and ocean resources to balance conflicting values; to working
with corporations to quantify the risks and opportunities of alternative resource
development options along their supply chains. Scaling from existing successes will rely
on three elements: codevelopment of new knowledge for decisions that demonstrate
interdependencies of ecosystems and human well-being; accessible approaches and tools
to replicate successes and continue learning; and building long-term capacity through
engaging leaders.*
. . . .

Protecting and restoring *natural capital*—Earth's lands, waters, and biodiversity, and the life-supporting goods and services that flow from these—have been promoted by many as the best hope for conserving biodiversity while also supporting sustainable, inclusive growth. Evidence is growing that illuminating the ways in which people depend on nature for their well-being can motivate efforts that improve both (Guerry et al. 2015; chapter 4, this volume). Creative yet highly dispersed innovation has shown glimpses of what is possible.

In this chapter, we outline how we might scale key advances in science, and policy, finance, and engagement mechanisms, to incorporate natural capital into resource- and land-use decisions and investment. Specifically, we summarize how practical, science-based approaches and tools can capture improvements in fundamental understanding of linkages among biodiversity, ecosystem services, and human well-being. These approaches and tools are being used to characterize the value and distribution of ecosystem services and integrate them into influential decisions.

Progress and Pathways to Scale

Mainstreaming biodiversity and ecosystem services into everyday decisions requires systematic methods for characterizing their value in currencies relevant to policy and finance. Unlike the case for traditional economic goods, where market prices can be used as a proxy for value, other methods are needed to quantify most of nature's values (MEA 2005; NRC 2005). Mainstreaming also requires policy or institutional reforms so that decision makers—policy leaders, managers, investors, and individuals—realize the full costs and benefits of their actions. Providing incentives for stewardship of natural capital, or disincentives for its depreciation, are needed to link short-term, individual interests with long-term societal well-being (Guerry et al. 2015).

Progress

Four major advances offer promise in face of mainstreaming needs. First, the Millennium Ecosystem Assessment, a visionary step in global science, was the first comprehensive assessment of the status and trends of the world's biodiversity and major ecosystem services. Its key finding—that two-thirds of our planet's ecosystem services were declining—captured the attention of world leaders (MEA 2005).

Second, the data and science linking ecosystem conditions and processes to human well-being are rapidly improving. Technological advances in remote sensing, computation, data science, and artificial intelligence are creating treasure troves of real-time data and enhanced ability to extract relevant information. Ecological science has become adept at mapping ecosystem services and the flow of benefits to people, from local to global scales (e.g., Guerry et al. 2015; Liu et al. 2015; Ouyang et al. 2016; Chaplin-Kramer et al., in prep.). Economic valuation methods have been applied to estimate the monetary value of benefits and their distribution to different segments of society (e.g., NRC 2005; Bateman et al. 2013). In addition, methods from other fields are now being applied in assessments and stakeholder processes to gain and incorporate better understanding of the psychological, social, and cultural importance of biodiversity and ecosystem services, and of shared values that people hold around nature (e.g., Pascual et al. 2017).

Third, experiments in payments for ecosystem services, in ecosystem-based management, investments in forest and other habitat restoration, and in regional- and national-level planning are proliferating. These illuminate which approaches offer the most promise (e.g., Arkema et al. 2015; Ouyang et al. 2016; Beatty et al. 2018; as well as chapters 9, 12, 13, and 15, this volume).

Fourth, there is encouraging momentum building around international measurement initiatives, in part driven by the need to assess progress toward achievement of the United Nations (UN) Sustainable Development Goals (SDGs). Across the public and private sectors, leaders are being encouraged and supported to integrate measures of natural capital stocks and flows into their decisions from local (community and business) to national scales. Led by international organizations such as the UN, the World Bank (e.g., Wealth Accounting and Valuation of Ecosystem Services Partnership–WAVES), and other multilateral institutions, and by business-led initiatives such as the Natural Capital Coalition, there is a rapidly growing network of measurement. Of note is the UN System of Environmental–Economic Accounting (UN–SEEA; UN et al. 2014), which has made great strides in harmonizing methodology and testing applications of natural capital accounts around the world. It is anticipated that at least 100 countries will have SEEA accounts by 2020. In addition, the Natural Capital Protocol (NCC 2016) is now widely embraced as the reference for integrating natural capital into business decision making.

Pathways

A wide range of strategies is needed to amplify natural capital solutions for in-clusive, green growth, and several pathways to scaling exist. As Lambin, Leape, and Lee (chapter 3) emphasize, scaling up from solutions proven at smaller scales involves implementation and amplification efforts from individuals to communities, from local to national governments, to investors and compa-nies with global supply chains. Durability of solutions is likely to be greater if businesses, governments, private investors, and individuals collectively are involved (see case 2, below).

One pathway to scaling is through *coproducing new science to incorporate natural capital values into policymaking and investments.* Science-based demonstrations are accelerating in institutions globally, driven by increased interaction between researchers and real-world generators and users of knowledge to enhance credibility, relevance, and impact of decisions.

These processes incorporate learning throughout in order to iteratively test, measure, and adapt natural capital approaches. Such engagements are informing diverse decisions related to spatial and climate adaptation planning on land and in coastal areas, payments for ecosystem services, impact assessments for permitting and mitigation, corporate risk mitigation, and habitat restoration (Guerry et al. 2015; Beatty et al. 2018; and chapter 16, this volume). Interdisciplinary science frontiers emerge in each case (e.g., Arkema et al. 2015; Ouyang et al. 2016; and chapter 9, this volume). Rigorous and ongoing evaluation of impacts of investments and policies on biophysical and human well-being metrics is an important next step along this pathway.

A second pathway is *incorporating new science into tools for promoting natural capital-based approaches.* Inspiring demonstrations and growing demand are driving development of accessible stakeholder engagement frameworks and tools. These are embedded within iterative decision-making processes including assessment, implementation, and adaptation. Drawing on new science captured in InVEST (Sharp et al. 2018) and kindred software, an explosion of tools is contributing to the growing ease with which practitioners and decision makers can estimate the values of ecosystem services to people, tailored for their specific needs (e.g., Guerry et al. 2015; Beatty et al. 2018; World Bank 2018). Much room for improvement remains, yet the impact is notable. The InVEST suite of tools has been codeveloped and tested with hundreds of researchers, practitioners, and managers, and it is the most widely used software for quantifying natural capital, with applications in over 160 countries (Posner et al. 2016). Countries, like China, have adopted them officially in development and conservation planning (Ouyang et al. 2016; Xu et al. 2017).

A third path is in *engaging leaders and practitioners to build capacity and magnify the impact of these successes.* The focus here is twofold: (1) on trainings in natural capital approaches and decision support software for building a community to mainstream natural capital systemically; and (2) on convening leaders from communities, governments, investors, and companies to share stories of success or ongoing impediments across high-leverage decision contexts. Engaging leaders in multisectoral decision processes can greatly encourage coordinated approaches to planning, mitigation, and investment to achieve positive outcomes and amplify meaningful change, as the demonstrations below illustrate.

Scaling Demonstrations

Growing evidence demonstrates how natural capital and other information are transforming policies and investments around the world. Development planning and compensation efforts by governments, and businesses' work to increase transparency and shift markets toward more sustainable commodity supply chains are promising examples (see case 2, below; and chapters 3, 9, 10, and 16, this volume). Here we highlight three illustrative cases among many.

Case 1. Scaling Climate-Resilient Coastal Development Investment in Latin America and the Caribbean

Coastal climate impacts through sea-level rise, warmer ocean temperatures, and changes in river runoff are exacerbating sociopolitical and environmental ills facing communities throughout Latin America and the Caribbean (LAC), with disproportionate effects on socioeconomically vulnerable groups. In this context, the Inter-American Development Bank (IDB) and the Natural Capital Project (NatCap) are cocreating ecosystem-service science and decision-support tools for coastal development planning with governments and stakeholders in the LAC region. The IDB is a leading multilateral investment institution for the region, supporting design and investment in coastal resilience, and sustainable and green infrastructure, including approximately US$2 billion annually of climate finance (IDB 2018a).

NatCap supported governments and communities to complete climate-smart, sustainable development plans—the Integrated Coastal Zone Management (ICZM) Plan in Belize (Clarke et al. 2016) and Andros Master Plan in The Bahamas (Office of the Prime Minister 2017; chapter 16, this volume). Both plans incorporate ecosystem service approaches to harmonize diverse stakeholder objectives around infrastructure, securing water and food supplies, risk management for coastal hazards, tourism benefits, and conserving coastal biodiversity such as coral reefs and mangrove forests. The plans are formally approved and being implemented through public and private investments, including those from the IDB. In Belize, the ICZM Plan contributed to the Belize Barrier Reef Reserve System (a UNESCO World Heritage Site) being removed from the list of "World Heritage at Risk" sites in 2018. Pilot implementation projects in Andros (The Bahamas) support communities and the private sector to develop livelihoods oriented toward sustainable uses of natural resources, such as birding- and fly-fishing-based tourism and sustainable supply chains for sponge and lobster harvest.

The science-based stakeholder engagement approaches and decision-support tools from these two demonstrations are ready to scale—and scaling is urgently needed. Accelerating investment in coastal infrastructure development, combined with risks from climate change, could undermine the natural resource-based economies throughout LAC if investments do not incorporate sustainability, natural capital, and resilience principles. To advance scaling, IDB and NatCap are aiding interested LAC governments through (1) increasing awareness of green policy options for increasing coastal resilience; (2) standardized guidance and methods for identifying and prioritizing green and hybrid (green+grey) infrastructure alternatives (e.g., Mandle, Griffin et al. 2016; IDB 2018b); (3) guidance for designing immediate "low regret" green solutions in a data-poor environment (while also improving information); and (4) technical training of practitioners in designing green infrastructure solutions, using open-source decision-support tools, and monitoring outcomes (e.g., Mandle, Douglass et al. 2016). In addition, the IDB has recently consolidated a Natural Capital Lab to drive innovation in conservation, landscape, biodiversity, and marine ecosystem finance, in both the public and the private sectors.

There is a key opportunity now to adapt and standardize natural capital development planning approaches and tools to incorporate (1) new data technologies (e.g., remote sensing, drones, mobile devices) for innovative applications (e.g., including specific beneficiaries such as city dwellers, smallholder farmers, indigenous peoples); and (2) policy and finance innovations for climate-smart development of cities, coastal zones, and nations. With improved technical capacity and increased awareness, a collaborative effort can amplify the opportunity in the LAC region to implement cross-sector, sustainable development planning, and connect to like-minded global efforts.

Case 2. Advancing a Sustainable Rangeland Economy in Mongolia

A major challenge across arid regions of the planet is transitioning to sustainable livelihoods in the face of a harsh and changing climate. These transitions can be especially daunting in rangeland systems where livestock productivity, ecosystem services, and quality of life for people are highly interconnected through rangeland condition. The case summarized here has implications far beyond Mongolia, as global actors in business, development finance, and civil society are engaging actively to innovate together, learn what is working, and then scale. Proving the case that new remote sensing technologies and ecosystem service analyses can inform both sustainable supply chain certification and evaluation of the International Finance Corporation's (IFC) performance standards for development impacts will help meet global verification demands that are critical for building confidence in approaches and resilience in outcomes.

Goat densities in Mongolia have increased fourfold over the past three decades, a response to increased global demand for cashmere, deregulation of the livestock sector, and the vagaries of harsh winters in a changing climate. Increased livestock density is wreaking havoc on people and wildlife—reducing financial certainty among herders, increasing dust storms, and affecting rangeland suitability for populations of wild ass (Khulan), goitered gazelle, and other rare wildlife. To help create a more sustainable and resilient rangeland system, a diverse group of interests is innovating together as the Sustainable Cashmere Project (SCP). SCP works with herders through cooperatives. The Luxury group *Kering* (which includes brands such as Gucci, Saint Laurent, and Balenciaga) is developing a sustainable source and supply chain of high-quality cashmere. The *Oyu Tolgoi* (OT) mining project (owned jointly by the Mongolian government and Rio Tinto) is applying the IFC performance standards to offset impacts from its large copper and gold mining operations, investing in herder livelihoods and healthy rangelands. The *Wildlife Conservation Society* oversees field monitoring for rangeland condition, promotes wildlife conservation, and engages with local nomadic herders to improve pasture management activities, technical capacity, and access to the developing sustainable cashmere supply chain.

The Natural Capital Project is helping the SCP track and verify integrated impacts of new financial and management interventions in the vast Gobi Desert. NatCap is advancing remote sensing technology and ecosystem services modeling to measure the effects of changes in grazing practices, thus augmenting sparse field sampling, to answer questions about the relationships between grazing density, cashmere quality, herders' income, rangeland conditions, and wildlife populations. Trust will grow if the herders—those most vulnerable during the shift to more sustainable grazing practices—see evidence of the modeled relationships between their actions, rangelands, herd size, and livelihoods. The year 2018 marks only the third season of SCP herders selling cashmere into the program, and the project continues to explore the best financial and social incentives to reward herders for their continued involvement.

The varied collaborators in the SCP have diverse motivations, but all care about healthier rangelands and more sustainable herder livelihoods. Aligning (1) market incentives through the developing sustainable cashmere market; (2) offset policies from OT mining, IFC financing standards, and the government of Mongolia; and (3) openly shared verification of rangeland ecosystem condition, wildlife status, grazing intensity and herder livelihoods, could lead to a more resilient system. The potential for scaling comes through sustainable global supply chains, application of IFC performance standards, and verification systems for supply chains and offset policies using remote sensing.

Case 3. Scaling Natural Capital Approaches through Gross Ecosystem Product (GEP) in China

China stands out globally in three key ways over recent decades: having the world's highest rates of economic growth and poverty alleviation; the highest rates of environmental degradation; and the great political will at the national level to reconcile this conflict. President Xi Jinping has seriously committed to reversing China's environmental degradation, declaring China's Dream of becoming the "ecological civilization of the 21ˢᵗ Century" through a transformation to inclusive green growth.

Placing ecological and economic information on a par, China is developing and adopting a new measure—Gross Ecosystem Product (GEP)—as a guide to securing sustainable improvements. GEP will help reveal social and economic contributions of ecosystems, highlight ecological connections among regions, guide financial compensation from ecosystem service beneficiaries to regions supplying those services, inform policy in green finance, and serve as a government performance metric. GEP is defined analogously to Gross Domestic Product (GDP): quantities of ecosystem goods and services are multiplied by prices using market and nonmarket valuation methods, and aggregated to obtain a tractable single measure that, over time, reflects the growth or decline of ecosystem services. And GEP (natural capital measure) can be reported alongside GDP (traditional economic performance measure) and the Human Development Index (HDI, focusing on health, education, and living standards) to track the three key dimensions of achieving China's Dream.

The initial phase of developing and testing a GEP approach was designed around four major elements, each drawing on pioneering work of the past two decades: (1) a system of natural capital accounts for tracking the magnitude and condition of biophysical stocks (Ouyang et al. 2016); (2) a systematic approach for translating these stocks into quantified flows of ecosystem services (Ouyang et al. 2016; Sharp et al. 2018); (3) a systematic approach for valuing ecosystem services; and (4) aggregating the values in (3) into a useful, single GEP metric comparable with GDP. These four steps together allow the development of GEP measured in value and physical terms.

As of October 2018, China was developing and testing GEP at the national level and in twenty-three counties, seven cities, and four provinces (Ouyang et al. 2017). Initial results show that GEP accounting can help evaluate conservation performance, the efforts of local governments, and the effectiveness of ecological compensation and related conservation policies. Efforts are underway to connect GEP with a UN-led ecosystem accounting project in China. There is potential for widespread adoption of this tractable measure of ecological performance for governments worldwide.

Lessons Learned

Mainstreaming natural capital into decisions is a long-term proposition, requiring coevolving advances in knowledge, social and governance institutions, and culture. Each new demonstration can contribute to the theory of change laid out here, with its three key elements: codevelopment of new science-based solutions to change decisions; tools to accelerate and scale new demonstrations; and building long-term capacity through engaging leaders. These elements for scaling are all within reach.

First, more examples of projects or enterprises that, as a result of properly valuing ecosystem services and natural capital, improve decisions, institutions, and human well-being, will instill confidence among leaders for scaling. Examples such as those presented here both test our knowledge against real-world problems and produce compelling stories of how an ecosystem services approach can make a difference.

Second, governments, businesses, and individuals must find it easy to integrate ecosystem services and natural capital into their decisions, and the methods for doing so must be transparent, credible, and predictable. Many sectors of society are open to the concepts, but do not know how to apply them in a tangible way.

Last, visible and charismatic successes will draw in political and thought leaders and thereby trigger broader awareness and attention. Building capacity in businesses, governments, communities, and investors will encourage mainstreaming of these approaches into decisions that are made every year and that impact our natural and social world.

Key References

Arkema, Katie K., Gregory M. Verutes, Spencer A. Wood, Chantalle Clarke-Samuels, Samir Rosado, Maritza Canto, Amy Rosenthal et al. 2015. "Embedding ecosystem services in coastal planning leads to better outcomes for people and nature." *Proceedings of the National Academy of Sciences* 112, no. 24:7390–95.

Bateman, Ian J., Amii R. Harwood, Georgina M. Mace, Robert T. Watson, David J. Abson, Barnaby Andrews, Amy Binner et al. 2013. "Bringing ecosystem services into economic decision-making: Land use in the United Kingdom." *Science* 341, no. 6141:45–50.

Beatty, C. R., L. Raes, A. L. Vogl, P. L. Hawthorne, M. Moraes, J. L. Saborio, and K. Meza Prado. 2018. *Landscapes, at Your Service: Applications of the Restoration Opportunities Optimization Tool (ROOT)*. Gland, Switzerland: IUCN.

Chaplin-Kramer, R., R. P. Sharp, C. Weil, E. M. Bennett, U. Pascual, A. L. Vogl, K. K. Arkema et al. Forthcoming. Global modeling of nature's contributions to people. *Science*.

Clarke, C., M. Canto, and S. Rosado 2016. *Belize Integrated Coastal Zone Management Plan*. Coastal Zone Management Authority and Institute (CZMAI).

Guerry, Anne D., Stephen Polasky, Jane Lubchenco, Rebecca Chaplin-Kramer, Gretchen C. Daily, Robert Griffin, Mary Ruckelshaus et al. 2015. "Natural capital and ecosystem services informing decisions: From promise to practice." *Proceedings of the National Academy of Sciences* 112, no. 24:7348–55.

Inter-American Development Bank (IDB). 2018a. *Sustainability Report 2017*. Inter-American Development Bank.

———. 2018b. *What Is Sustainable Infrastructure? A Framework to Guide Sustainability across the Project Cycle*. Technical Note IDB-TN-1388.

Liu, Jianguo, Harold Mooney, Vanessa Hull, Steven J. Davis, Joanne Gaskell, Thomas Hertel, Jane Lubchenco et al. 2015. "Systems integration for global sustainability." *Science* 347, no. 6225:1258832.

Mandle, L., R. Griffin, J. Goldstein, R. Acevedo-Daunas, A. Camhi, M. Lemay, E. Rauer, V. Peterson. 2016. *Natural Capital & Roads: Managing Dependencies and Impacts on Ecosystem Services for Sustainable Road Investments*. Monograph del BID: 476.

Mandle, Lisa, James Douglass, Juan Sebastian Lozano, Richard P. Sharp, Adrian L. Vogl, Douglas Denu, Thomas Walschburger, and Heather Tallis. 2016. "OPAL: An open-source software tool for integrating biodiversity and ecosystem services into impact assessment and mitigation decisions." *Environmental Modelling & Software* 84:121–33.

Millenium Ecosystem Assessment (MEA). 2005. *Ecosystems and Human Well-being: Synthesis*. Washington, DC: Island Press.

NCC (Natural Capital Coalition). 2016. "Natural capital protocol." www.naturalcapitalcoalition .org/protocol.

NRC (National Research Council). 2005. *Valuing Ecosystem Services*. Washington, DC: National Academies Press.

Office of the Prime Minister of The Bahamas. 2017. "Sustainable Development Master Plan for Andros Island." http://www.vision2040bahamas.org/media/uploads/andros_master_plan .pdf.

Ouyang, Z., L. Jin. 2017. *Developing Gross Ecosystem Product and Ecological Asset Accounting for Eco-Compensation*. Beijing: Science Press.

Ouyang, Zhiyun, Hua Zheng, Yi Xiao, Stephen Polasky, Jianguo Liu, Weihua Xu, Qiao Wang et al. 2016. "Improvements in ecosystem services from investments in natural capital." *Science* 352, no. 6292:1455–59.

Pascual, Unai, Patricia Balvanera, Sandra Díaz, György Pataki, Eva Roth, Marie Stenseke, Robert T. Watson et al. 2017. "Valuing nature's contributions to people: The IPBES approach." *Current Opinion in Environmental Sustainability* 26:7–16.

Posner, Stephen M., Emily McKenzie, and Taylor H. Ricketts. 2016. "Policy impacts of ecosystem services knowledge." *Proceedings of the National Academy of Sciences* 113, no. 7:1760–65.

Sharp R., R. Chaplin-Kramer, S. Wood, A. Guerry, H. Tallis, T. Ricketts et al. 2018. *InVEST User Guide*. http://data.naturalcapitalproject.org/nightly-build/invest-users-guide/html/.

United Nations, European Commission, Food and Agriculture Organization of the United Nations, International Monetary Fund, Organisation for Economic Co-Operation and Development, and World Bank. 2014. *System of Environmental-Economic Accounting 2012 Central Framework*. New York: United Nations. https://seea.un.org/sites/seea.un.org/files /seea_cf_final_en.pdf.

World Bank. 2019. *Evaluation of Approaches to Global Ecosystem Services Modeling*. R. Chaplin-Kramer, L. Mandle, and J. Johnson, eds. World Bank Report.

Xu, Weihua, Yi Xiao, Jingjing Zhang, Wu Yang, Lu Zhang, Vanessa Hull, Zhi Wang et al. 2017. "Strengthening protected areas for biodiversity and ecosystem services in China." *Proceedings of the National Academy of Sciences* 114, no. 7:1601–6.

CHAPTER 3

Amplifying Small Solutions for Systemwide Change

Eric F. Lambin, Jim Leape, and Kai Lee

A major challenge is the design and implementation of solutions that lead to transformations at the scale of a large, complex problem. The polycentric model of governance recognizes the importance of multiple vectors in driving change. Here we argue that multiple small, bottom-up solutions can have synergistic effects and reinforce each other to lead to rapid change provided that, in addition to the design of small solutions, efforts are dedicated to their up-scaling. The amplification mechanisms that create tipping points of adoption of small solutions include (1) changes in social norms that influence consumer behaviors, (2) public policies that support small solutions or make them mandatory, and/or (3) commitments by dominant private companies that apply solutions across their value chains and their sector. This theory of change for sustainability makes small actors feel empowered, as small solutions implemented locally, within the scale of competence of each actor, can then be scaled up and lead to systemwide change.
....

Biodiversity loss, the conversion of natural ecosystems, the impact of climate change on ecosystems, and the depletion of fish stocks in the oceans are all complex, "wicked" problems that lack an easily identifiable and generalizable solution (DeFries and Nagendra 2017). Most of these problems are large in scope as they cover multiple geographies, sectors, and stakeholders. They also involve collective action problems: individuals make independent decisions about costly actions, but outcomes of these decisions benefit or harm an entire group. As a result, cooperation would lead to a better outcome for all, but individuals fail to cooperate because of conflicting interests that discourage joint action.

A major challenge with this class of wicked problems is to design and implement solutions that are appropriate for each context and yet lead to transformations at the scale of the problem (DeFries and Nagendra 2017). Many solutions address only a fraction of the problem and lack potential to achieve large-scale impact. In some cases, local interventions lead to leakage: success in one place or sector displaces unsustainable behaviors in other places or sectors. For example, countries that protect their forests simultaneously increase their imports of timber and thus offshore their deforestation (Meyfroidt et al. 2010). Leakage reduces or even negates the already small benefits obtained.

The most common approach to such large-scale, wicked problems is to assume that the scale of the solution needs to match the scale of the problem and therefore to focus on designing large-scale solutions. The UNFCCC's efforts to solve climate change through a multilateral agreement involving all countries and many observers exemplifies this approach. Meaningful commitments that cover all countries and sectors of activities take years to be negotiated, are fraught with institutional complexities, and suffer from implementation issues, such as the need for many exceptions and adaptations to local conditions, the lack of sanctions in case of noncompliance, and the possibility of withdrawing from the agreement.

At the other extreme, many efforts set out to design small-scale, local solutions to a large problem in hopes that they will be so attractive that many others will adopt them and that, cumulatively, they will lead to a meaningful transformation of the large-scale system. This approach was illustrated in the French documentary *Demain* (*Tomorrow*), a box office hit, which showcased simple local-scale solutions to sustainability challenges around the world, such as urban vegetable gardens, zero-waste initiatives at the scale of a plant or city, local renewable energy projects, and so forth. This approach is attractive because these small solutions are easy to implement at the levels of individuals, communities, or local political powers, which has been referred to as acting "within the scale of our competence" (Berry 1991). But small solutions often stay small. A recurrent, vexing question is how to promote wide and rapid adoption of small solutions, beyond the small fraction of a population that already cares about sustainability.

Most successes have come from a combination of highly scalable small-scale solutions with mechanisms that amplify their implementation and impact, leading to their up-scaling. This pairing may happen by design or opportunistically. In this approach, multiple actors who may or may not coordinate their actions contribute to an up-scaling of the most successful of these solutions once these have gained traction locally. This occurs for example when a sustainable production standard that was adopted voluntarily by a small fraction of producers gets integrated into a mandatory public policy or in the sourcing practice of a large company, thus impacting all producers in a jurisdiction or supply chain. Another example is the up-scaling of individual decisions to commute by bike, which is achieved by new policies (e.g., congestion charges), land use planning, and investments in safe cycling infrastructure that create a tipping point in urban mobility. Since 2016 in Copenhagen, for example, there are more bikes than cars on the streets. In the same way, a tipping point in adoption of electric cars can be created by financial incentives and public and private investments in charging infrastructure. China, which accounted for half of global electric car sales in 2017, exempts electric cars from purchase taxes since 2014 and has an ambitious program to install new charging stations and poles across the country. This approach thus includes two components: small-scale solutions that are scalable and attractive, and amplification mechanisms that bring them to scale.

The idea that big problems are best solved by multiple small, bottom-up solutions that are then amplified, rather than by large, top-down solutions, is gaining traction. Instead of focusing only on global efforts, scholars are calling for polycentric approaches that achieve benefits at multiple scales (Ostrom 2010). A governance arrangement is polycentric when there are many centers of decision making at differing scales that are formally independent from each other. Extensive research on polycentric governance shows that small- and medium-scale units are important components for the management of large-scale problems, as small solutions have the potential to reinforce each other. Given the complexity of social change, a polycentric approach to solutions can foster a profusion of experiments and conditions for scaling and see what takes off. By contrast, top-down solutions assume one can figure out the answer and prescribe it.

In sum, there is an increasing recognition that (1) small solutions are easier to implement than larger ones, and (2) some successful small solutions can be scaled up or catalyze large-scale change. Below, we examine these two propositions, with a focus on the nexus between public sector and market changes.

Small Solutions to Big Problems

The Chinese philosopher Lao Tzu wrote in the sixth century BC that "a journey of a thousand miles begins with a single step." Solutions can be defined as small if their scope is very limited compared to the scale of the problem they attempt to address. A solution is small if it is implemented only by individual consumers, companies, communities, or governments in a bottom-up approach. For example, when efforts to forge a global convention to address forest degradation failed at the Rio Conference in 1992, conservationists turned to a small solution: enlisting individual companies or forest landowners to shift to sustainable production. The creation of the Forest Stewardship Council, providing a system for certification and labeling of products from sustainably managed forests, created a vehicle that founders hoped would help take those small solutions to scale.

Small solutions are attractive because many large-scale problems are the result of the cumulative impact of small actions taken by individuals, households, communities, private firms, and governments. Small solutions can target the root-cause drivers of the problem. Small solutions are easier to implement than large ones for several reasons. Small solutions are generally better understood by stakeholders than big ones as they tend to be simpler, clearer, and more visible (Heath and Heath 2010). There can be less resistance to implementing small solutions because they are within reach, which leads to higher rates of adoption if they are voluntary or to better compliance if they are mandatory. Small solutions are implemented in one place or sector among stakeholders who are part of the same social network and therefore can enjoy a higher level of trust with one another. With easier communications among stakeholders, a local solution is more readily adapted to the specific context. Small solutions do not require global consensus or even collective decision making: they can be implemented by a single actor or by "coalitions of the willing" (Hawken 2008). Finally, monitoring of the implementation of small solutions and their enforcement is less costly than for large, complex ones. The above analysis underlies the slogan "Think globally but act locally."

Scaling Up Small Solutions

The theory of innovation diffusion describes the process by which a new idea, practice, or object is communicated through certain channels over time among the members of a social system (Rogers 2003). Diffusion is thus viewed as a kind of social change, with a potential to be amplified.

The Theory of Innovation Diffusion

Adoption of innovations is nonlinear: there are inflection points corresponding to thresholds of adoption and compliance. According to diffusion theory, the spread of an innovation within a region or sector can be represented by a logistic (or S-shaped) curve (Rogers 2003). The proportion of possible users of the innovation who are adopters increases slowly at first, then rapidly, and then slowly again as saturation is reached. Up-scaling adoption means getting an innovation on to this curve (i.e., getting it to take off), advancing the first inflection point when adoption accelerates and raising the saturation point of adoption, so that it reaches a higher level of diffusion before leveling off.

Potential users of an innovation may only adopt it after the relevant information has reached them, after their resistance to change has been overcome, and once they have acquired the necessary know-how (Casetti 1969). Rogers (2003) explains that the rate of adoption of an innovation depends on its perceived attributes: its relative advantage (is it perceived as better than the solutions it supersedes?); its compatibility (is it perceived as consistent with existing values and needs of potential adopters?); its complexity (is it difficult to understand and use?); its trialability (can it be experimented with on a limited basis?); and its observability (are the results of the innovation visible to others?). A common narrative is that an innovation spreads as new users hear of its benefits, then learn how to use the innovation in their own activities. Their success spreads the news to others, accelerating adoptions until there are few remaining holdouts, and the innovation cycle slows.

These processes of change have mostly been studied at the individual or firm levels. Several examples illustrate how new consumer behaviors eventually reach a tipping point of adoption once they shift from being marginal to becoming new social norms, for example, quitting smoking, shifting to a more vegetarian diet, or acquiring an electric car (Nyborg et al. 2016).

Scalable Solutions

Some small solutions have a greater potential to be scalable than others. Scalability depends first on whether a solution is attractive, for example, because it solves an important problem, it is credible with key stakeholders, or it will be profitable. The Better Cotton Initiative, for example, has grown rapidly in part because, in introducing techniques that reduced farmers' use of water and chemical inputs (two major environmental problems), it also reduced their costs and increased their profits.

Scalability also depends on whether a solution is highly adaptable and thus can easily be applicable to different contexts. Initially, the implementation of a small solution acts as a proof-of-concept and is used to test its feasibility, improve its design, and demonstrate its effectiveness and efficiency. This requires rapid learning from initial failures and iterative adjustments as the solution is being applied in different environments.

Mechanisms for Getting to Scale

A sustainability transition is characterized by strong and evolving interactions among a wide array of players, which include diverse nongovernmental organizations (NGOs), companies, national and subnational governments, and international organizations. Several elements accelerate systemwide transformations.

Salience

Change starts with salience: a recognition among key actors that there is a problem and that it is in their interest to act. Transparency, based on the collection, disclosure, and wide dissemination of data and information on actors' practices, can create that kind of salience, spurring adoption of new practices. Information may be a surprisingly strong motivator to change behaviors as actors try to avoid bad publicity and preserve their reputation (Thaler and Sunstein 2008). In addition to the principle that one manages only what is measured, actors are made aware of risks associated with unsustainable practices. In China, for example, the Institute for Public & Environmental Affairs famously created the China Water Pollution Map, a public database of information on pollution by factories in China, which spurred action both by government enforcement agencies and by the factories' customers. Thus, transparency creates potential for public or private actors to spot and manage problems or risks (Gardner et al. 2018). In a more transparent world, adoption of sustainable sourcing practices can play an important role in preempting disaster or responding to crises, as illustrated by the response of the apparel industry to the Rana Plaza accident, or the response of food and agriculture companies to deforestation in the Amazon.

Often NGOs have played an important role in creating transparency—bringing issues to public attention, generating pressure for governments or companies to address them, and stimulating accountability. On local issues such as water scarcity or mining practices, pressure from local community NGOs has forced companies to take action to protect their social license to operate. International campaigns have held global companies more broadly accountable for impacts in their supply chains, prompting them to act to protect their brands.

First Movers

Often it is smaller governments or companies that are first to act, but they can play a crucial role—proving the practicality of a solution, and inspiring others to follow. At the country level, since the 1980s, the government of Costa Rica has been a pioneer in forest protection and in creating the first national-scale scheme of payments for ecosystem services (PES). Arguably, the success of Costa Rica's program has inspired the design of some of the national and international implementations of REDD+, which are efforts by many countries to reduce emissions from deforestation and forest degradation and foster conservation, sustainable management of forests, and enhancement of forest carbon stocks. REDD+ creates a financial value for the carbon stored in forests.

In the private sector, early adopters also help to build a market for sustainable products, shift the norms and expectations among consumers, and thus create competitive pressure. For example, progressive retailers—such as Whole Foods in the United States, Sainsbury's in the United Kingdom, Woolworths in South Africa, or Migros in Switzerland—built their businesses in part on a commitment to sustainability. They proved the viability of that approach. They also shifted expectations of their customers, who were seen by other retailers as a highly visible and desirable group to attract, and of consumers more broadly.

Big Players

To get to scale, a solution has to be adopted by big players. In market-based efforts, once early adopters have paved the way, large consumer-facing companies—retailers and consumer products companies—have often been the next to move. Such companies are particularly attuned to public concerns and exquisitely sensitive to risks to their valuable brands. Thus, after Whole Foods, Sainsbury's and other early adopters had pioneered sourcing of sustainable seafood certified by the Marine Stewardship Council (MSC), Walmart announced that it would also shift its purchasing and sparked a surge in fisheries seeking certification, creating a clear inflexion point in growth of MSC's market share. As this example shows, consumer-facing companies can play a crucial role in bringing in the traders and producers that supply them. As a growing number of consumer companies committed to eliminating deforestation from their supply chains, for example, they were actively engaged in inspiring Cargill, Wilmar, and other key traders and growers to make similar commitments.

For decades, California—which represents about 12 percent of the US population and is the fifth largest economy in the world—has been a big player in environmental protection. California's initiative on air quality in the 1950s established a reputation for leadership on environmental standards. In the 1970s, for example, California pioneered energy efficiency standards for home appliances such as refrigerators that proved to be both effective and economical, and were subsequently adopted by federal regulators. California's adoption of these innovations created the political opening and impetus for federal and state governments to act, while also creating momentum in the marketplace, as manufacturers complied with the California standards.

Dense Social Networks

Dense social networks (i.e., with many connections) also facilitate the diffusion of innovations and peer pressure for adoption (Rogers 2003). Social pressure within a community that is characterized by a dense social network strengthens adoption of, and compliance with, small solutions. Links among social networks can then be a path to scale. This can occur through networks of actors within sectors, such as the C40 for mayors or the Consumer Goods Forum for companies. There are also important networks that link actors across sectors, such as the Sustainable Seafood Summit, the Tropical Forest Alliance, or the World Economic Forum. In the private sector, some of these platforms support precompetitive collaborations on sustainability. Note that dense social networks can also be obstacles to diffusion when they defend entrenched interests and the lowest common denominator—as in the case of many trade associations, the US Chamber of Commerce, or the European Round Table.

Supply chains are also networks of tightly linked economic actors, which can act as vehicles for adoption of new practices and standards. As noted above, an economically dominant actor in a supply chain, who has strong motivations for moving toward sustainability can impose sustainability standards that affect millions of producers in some cases (Thorlakson et al. 2018).

Leadership

Leadership is key in the up-scaling of good practices given the role of charismatic leaders as change agents, especially when action is required across supply chains and across sectors. Corporate leaders like Paul Polman of Unilever, Yves Chouinard of Patagonia, and Feike Sijbesma of DSM have played this role in the adoption of sustainability standards in their sectors. They have been role models for other CEOs, have engaged their companies' suppliers and customers, and have forged partnerships with governments, civil society, and even their competitors. Mohammed Yunnus was a similarly charismatic champion for microcredit, taking the experience of the Grameen Bank to many other countries.

Right Time
Implementing small solutions at the most opportune moment is also critical to their successful up-scaling. This is captured by the ancient Greek concept of "Kairos" that signifies the proper time for action. A small solution is much more likely to have a large-scale impact if it is introduced at the right time, when there is an opening and receptivity for a new solution by potential adopters. For example, the growth of organic agriculture in Europe coincided with well-publicized food contamination scandals, which had created a demand among consumers for food that is free of chemicals and whose origin is traceable.

Policy Interactions Facilitate Up-Scaling

In some circumstances, small solutions undertaken by diverse actors can cata-lyze and leverage each other, producing an effect that is more than additive. The rich interplay between government, NGOs, and private company poli-cies is key to the up-scaling of interventions to promote sustainable practices (Lambin and Thorlakson 2018). Various policies from different decision cen-ters can be complementary, collaborative, or coordinated, or they can sub-stitute for, supersede, or coopt each other. Public and private environmental governance regimes rarely operate independently and tend to reinforce each other in ways that can promote (or hinder in some cases) sustainability (Lam-bin and Thorlakson 2018).

These complementarities greatly facilitate scaling up interventions aimed at promoting sustainability. An example has been the emergence of certifica-tion schemes. The history of these voluntary sustainability standards illustrates how interactions between private and public policies scales up adoption of sustainable practices. Increasingly, the uptake of voluntary sustainable stan-dards developed by NGOs or multi-stakeholder initiatives is significantly amplified when these standards are adopted by governments in their environ-mental policies. For example, the legal criteria for sustainable forest manage-ment in Bolivia's forest code are copied from the Forest Stewardship Council (FSC) standard (Ebeling and Yasue 2009). In Brazil, the "Certifica Minas Café" certification standard, implemented by the government of the State of Minas

Gerais was developed based on the Utz's standard for coffee (Glasbergen and Schouten 2015).

Under jurisdictional approaches to certification, whereby adherence to sustainability criteria is required for an entire jurisdiction, governments formally endorse (and impose) standards designed by NGOs or commodity roundtables. For example, the governments of Ecuador, the State of Sabah (Malaysia), and the province of Central Kalimantan (Indonesia) are implementing a jurisdictional approach to Roundtable for Sustainable Palm Oil (RSPO) certification of palm oil production (Lambin and Thorlakson 2018). In other cases, importing countries can provide legitimacy to certification schemes and stimulate their market uptake. For example, the European Union's Forest Law Enforcement, Governance and Trade (FLEGT) action plan accepts the FSC and Programme for the Endorsement of Forest Certification (PEFC) certifications as evidence of legal and sustainable timber from all regions and producer countries (Gulbrandsen 2014). The US Lacey Act also includes a public recognition of private certification schemes for timber, as providing evidence of "due care" (Overdevest and Zeitlin 2014). Public sector initiatives can thus act as platforms for dissemination of new, sustainable practices by "socializing" ideas and building legitimacy.

Companies also contribute to up-scaling sustainability standards when these companies commit to sourcing or producing only certified products. Sustainability standards are then integrated as part of companies' internal codes of conduct. They are therefore applied to the entire operations and supply chains of these companies. Many firms and supply chains can collaborate in using a sustainability certification scheme, enabling economies of scale in the certification process and strengthening its credibility. In 2015, more than 85 percent of companies with commitments to reduce deforestation in the palm, timber and pulp, soy, and cattle supply chains relied on third party certification to identify commitment-compliant commodity supply (Peters-Stanley et al. 2015; Lambin, Gibbs et al. 2018). As other examples of corporate actors adopting NGO-led certifications to implement their aspirational pledges, Starbucks became the largest US buyer of Fair Trade coffee, and IKEA's sustainability commitments rely on certifications such as FSC and Better Cotton Initiative (BCI). Many companies such as Walmart or Starbucks also rely on partnerships with NGOs to develop and implement their corporate sustainability programs (Elder and Dauvergne 2015).

The history of bikeshare programs provides another example of how the interplay between public and private actors facilitated the up-scaling of a small solution (https://www.citylab.com/city-makers-connections/bike-share). The first bikeshare scheme was introduced in Amsterdam by a group of activists in 1965. Plagued by thefts, it quickly collapsed. It wasn't until thirty years later that the idea was resurrected in Copenhagen by two entrepreneurs. In the years that followed, boosted by the invention of antitheft systems based on magnetic cards, bikeshares began to take off. Creation by Paris' city authorities of Velib, the city's ambitious bikeshare program, was an inflection point—there had been 75 bikeshare programs created in the previous twelve years; more than 1,600 were launched in the decade that followed (Economist 2017). Bikeshares spread across Europe first, and then North America, Asia, and the rest of the world. The private sector responded to this surge in interest, creating new technology for sharing systems, and, in some cities, authorities delegated the management of bikeshare services to private companies. Start-ups pioneered innovations: MoBike in China, among others, developed bikeshare systems without stations—customers could leave a bike wherever they wanted and a subsequent user would pick it up. It is estimated that today there are bikeshare systems in a thousand cities around the world. As of 2017, there were an estimated 16 million bikes in bikeshare systems in China alone.

Conclusion

With appropriate amplification mechanisms, small solutions can trigger a cascade of events leading to large-scale change. These cascading effects occur through networks of similar stakeholders—companies or government agencies—who imitate each other, compete for reputation, or enter into precompetitive collaborations to move together. These networks can also include dissimilar stakeholders—NGOs, businesses, and governments—who act synergistically, as in the case of certification for sustainable production or in the adoption of public policies that lock in sustainability improvements in economic activities. In some cases, independent actions by multiple actors may converge powerfully at random moments. In other cases, actors collaborate to address a problem. Both cases correspond to a polycentric model of governance whereby multiple strands of solutions, however small they are at the outset, build on each other and synergistically create a virtuous spiral toward change.

Key References

Berry, Wendell. 1991. "The futility of global thinking." In *Learning to Listen to the Land*, Bill Willers, ed., 150–56. Washington, DC: Island Press.

Casseti, Emilio. 1969. "Why do diffusion processes conform to logistic trends?" *Geographical Analysis* 1:101–10.

DeFries, Ruth, and Harini Nagendra. 2017. "Ecosystem management as a wicked problem." *Science* 356:265–70.

Ebeling, Johannes, and Maï Yasue. 2009. "The effectiveness of market-based conservation in the tropics: Forest certification in Ecuador and Bolivia." *Journal of Environmental Management* 90:1145–53.

Economist. 2017. "How bike-sharing conquered the world." *Economist*. December. https://www.economist.com/christmas-specials/2017/12/19/how-bike-sharing-conquered-the-world.

Elder, Sara D., and Peter Dauvergne. 2015. "Farming for Walmart: The politics of corporate control and responsibility in the global South." *Journal of Peasant Studies* 42:1029–46.

Gardner, Toby A., Magnus Benzie, Jan Börner, Elena Dawkins, Steve Fick, Rachael Garrett, Javier Godar, Andreanne Grimard, Sarah Lake, Rasmus K. Larsen et al. 2018. "Transparency and sustainability in global commodity supply chains." *World Development*. https://doi.org/10.1016/j.worlddev.2018.05.025.

Glasbergen, Peter, and Greetje Schouten. 2015. "Transformative capacities of global private sustainability standards." *Journal of Corporate Citizenship* 58:85–101.

Gulbrandsen, Lars H. 2014. "Dynamic governance interactions: Evolutionary effects of state responses to non-state certification programs." *Regulation & Governance* 8:74–92.

Heath, Chip, and Dan Heath. 2010. *Switch: How to Change Things When Change Is Hard*. New York: Crown Business.

Hawken, Paul. 2008. *Blessed Unrest: How the Largest Social Movement in History Is Restoring Grace, Justice, and Beauty to the World*. New York: Penguin Books.

Lambin, Eric F., Holly K. Gibbs, Robert Heilmayr, Kimberly M. Carlson, Leonardo Fleck, Rachael Garrett, Yann le Polain de Waroux, Constance L. McDermott, David McLaughlin, Peter Newton et al. 2018. "The role of supply-chain initiatives in reducing deforestation." *Nature Climate Change* 8:109–16.

Lambin, Eric F., and Tannis Thorlakson. 2018. "Sustainability standards: Interactions between private actors, civil society and governments." *Annual Review of Environment and Resources* 43. doi.org/10.1146/annurev-environ-102017-025931.

Meyfroidt, Patrick, Tom K. Rudel, and Eric F. Lambin. 2010. "Forest transitions, trade, and the global displacement of land use." *Proceedings of the National Academy of Sciences* 107 (49): 20917–22.

Nyborg, Karine, John M. Anderies, Astrid Dannenberg, Therese Lindahl, Caroline Schill, Maja Schlüter, Neil W. Adger, Kenneth J. Arrow, Scott Barrett, Stephen Carpenter et al. 2016. "Social norms as solutions." *Science* 354 (6308): 42–43.

Ostrom, Elinor. 2010. "Polycentric systems for coping with collective action and global environmental change." *Global Environmental Change* 20:550–57.

Overdevest, Christine, and Jonathan Zeitlin. 2014. "Assembling an experimentalist regime: Transnational governance interactions in the forest sector." *Regulation & Governance* 8 (1):22–48

Peters-Stanley, Molly, Stephen Donofrio, and Ben McCarthy. 2015. *Supply-change: Corporations, Commodities, and Commitments That Count*. Washington, DC: Forest Trends.

Rogers, Everett. 2003. *Diffusion of Innovations.* 5th ed. New York: Free Press of Glencoe.

Thaler, Richard H., and Cass R. Sunstein. 2008. *Nudge. Improving Decisions about Health, Wealth, and Happiness.* New Haven and London: Yale University Press.

Thorlakson, Tannis, Joann de Zegher, and Eric F. Lambin. 2018. "Companies' contribution to sustainability through global supply chains." *Proceedings of the National Academy of Sciences,* 115 (9):2072–77.

CHAPTER 4

Collaborative Approaches to Biosphere Stewardship

Carl Folke, Beatrice E. Crona, Victor Galaz, Line J. Gordon, Lisen Schultz, and Henrik Österblom

In what way can human actions and actors be redirected to appreciate the significance of natural capital, reconnect development to the biosphere, and help stabilize Earth in conditions favorable for a globalized, human-dominated world? Here we present seven examples of transdisciplinary collaboration aimed at transforming development toward pathways of sustainability and biosphere stewardship. They represent ongoing experiments founded in resilience thinking, and reflect science for change in cooperation with stakeholders in local places to international bodies and transnational actors. They are concerned with stewardship of natural capital, biodiversity, and ecosystem services in diverse settings, and with diverse practices, worldviews, and values covering issues like human well-being, livelihoods, and poverty alleviation. The collaborations have in common the view of humans and nature as intertwined social-ecological systems, embracing complexity, enabling inclusive decision making, promoting flexibility and learning, and leveraging opportunity and innovation.

The world is a complex moving target. The human dimension has accelerated into a major force driving the dynamics of the Earth system and, as it seems, into unfamiliar terrain beyond the glacial-interglacial dynamics of the last 1.2 million years (Steffen et al. 2018). There is definitely something new under the sun!

Characteristics of the new situation include intertwined social and ecological systems, cross-scale interactions of the intertwined systems, as well as complex dynamics with systemic tipping points and true uncertainty. The current speed, connectivity, and extent of human actions link previously unconnected sectors and ecosystems and generate new dynamics that can result in cascading change across countries and regions affecting human well-being and essential ecosystem services. Local events can escalate into global challenges, and local places are shaped by global dynamics (Reyers et al. 2018).

A major challenge in this new situation is to find ways to transform human actions toward pathways of sustainability that (1) appreciate the significance of natural capital, (2) reconnect development to the biosphere, and (3) help stabilize Earth in conditions favorable for a globalized, human-dominated world.

It is in this context that we approach the challenge of active stewardship of natural capital and the magnificent biosphere, including its biodiversity. Biosphere stewardship makes clear that we as humans are embedded within the biosphere. Sustainability requires improved stewardship of human actions in concert with the biosphere, from the local to the global and across scales, because the capacity of the biosphere serves as the foundation upon which the success of human futures rests. Stabilizing the Earth system in favorable conditions requires building resilience for complexity and change (Folke et al. 2016).

In this chapter, we present seven ongoing efforts of transformation toward biosphere stewardship in the interface of science-practice-policy-business. These efforts, founded on resilience thinking, reflect transdisciplinary platforms and arenas of collaboration for change and coproduction of knowledge for action. Science serves a significant function in all these efforts as an objective, nonpartisan knowledge provider.

Science for Collaborative Change

Biosphere stewardship is not a top-down global approach enforced on people, nor solely a bottom-up approach. It is a process engaging people to collaborate across levels and scales and with shared visions and creativity framed by proper institutions, continuously learning and gaining experience and building capacity to live with change, adapt, and transform. It could be viewed as an emerging ethic that embodies the responsible relationship of humans as part of and dependent on the biosphere.

Transdisciplinary science, where science contributes as an independent actor and interacts and coproduces knowledge with practice, policy, and business, plays a central role in the following examples aiming for biosphere stewardship.

Place-based Adaptive Governance for Transformation

While biosphere stewardship entails cross-scale collaboration, it is ultimately place-based, emerging from direct interactions between people and the living world. Many of our case studies illustrate how place-based governance, that is, governance that is informed by and adaptive to local places in a cross-scale and even global context, can support transformations toward biosphere stewardship.

For example, our collaborations with and comparison of local adaptive governance systems in Kristianstads Vattenrike (Sweden), regional adaptive governance of the Great Barrier Reef (Australia), and global adaptive governance of the Southern Ocean fisheries have identified three common features toward biosphere stewardship: (1) building knowledge and awareness of natural capital dynamics, allowing actors to learn and respond in an informed manner; (2) drawing on the diverse competences of state and nonstate actors, accessing a wider range of governance tools and approaches; and (3) enabling coordination, negotiation, and collaboration across whole landscapes and seascapes, allowing issues to be addressed in a holistic manner at the appropriate scale (Schultz et al. 2015).

Another prominent example of place-based governance systems, involving a range of actors including scientists, is the UNESCO biosphere reserves, which aim to maintain and develop ecological and cultural diversity and to secure ecosystem services for human well-being. The World Network of Biosphere Reserves currently includes 686 sites in 122 countries, some of which have been operating since the mid-1970s. Drawing on 177 interviews in 11 of the sites, we found that the biosphere reserve concept truly comes alive and shapes action when local meanings were connected to the global framework. In Kruger to Canyons (South Africa), the biosphere reserve was described as a "connector" in a landscape shaped by apartheid, whereas in wealthy Noosa (Australia), it was seen as a platform to celebrate sense-of-place (Schultz et al. 2018).

A place-neutral governance system misses the opportunity of drawing on and enhancing the engagement that comes from local identity and knowledge for transformations toward biosphere stewardship (Marshall et al. 2012).

Seeds of a Good Anthropocene

Seeds are initiatives (social, technological, economic, or social-ecological ways of thinking or doing) that exist at a minimum in prototype form. They represent a diversity of practices, worldviews, values, and regions, but are not currently dominant or prominent in the world. Bennett et al. (2016) collected examples of "seeds of a good Anthropocene," as well as critical information about them, using an international participatory process that helped identify and understand components and processes, which may lead to the emergence and scaling of initiatives fundamentally changing social-ecological relationships.

Through an analysis of 100 identified seeds, existing practices were discovered that offer the opportunity to (1) understand the values and features constituting a good Anthropocene; (2) determine the processes that lead to the emergence and growth of initiatives that fundamentally change social-ecological relationships; and (3) generate creative, bottom-up scenarios that feature well-articulated pathways toward a more positive future. It was found that inspirational visions can be key components of transformations to sustainability by shaping the very reality they forecast or explain.

The seeds project, transdisciplinary in scope and process, represents a novel approach to envisioning the future. The approach has the potential to help accelerate the adoption of pathways to transformative change for biosphere stewardship. The project is part of the Programme of Ecosystem Change and Society (PECS), a follow-up effort of the subglobal assessment of the Millennium Ecosystem Assessment and a core project of Future Earth. PECS aims to integrate research on the stewardship of social-ecological systems and the relationships among natural capital, human well-being, livelihoods, inequality, and poverty. PECS emphasizes that ecosystem services are not generated by ecosystems alone, but by social-ecological systems, in interacting local to global contexts.

Multiple-Evidence Base, the Convention on Biological Diversity and the Intergovernmental Science-Policy Platform on Biodiversity and Ecosystem Services

In complex adaptive systems, no single actor has the full picture, and making sense of the situation requires continuous learning, drawing on all available sources of knowledge, including local, indigenous, and scientific knowledge. But in practice, the integration of different kinds of knowledge has proven challenging, given the different systems of validation and asymmetric power relations between different kinds of knowledge holders. We have experienced such tensions in the context of the Convention on Biological Diversity (CBD) as well as the Intergovernmental Science-Policy Platform on Biodiversity and

Ecosystem Services (IPBES). For this purpose, the Multiple-Evidence Base has been developed. This approach to biosphere stewardship accepts that diverse knowledge systems are incommensurable, and that asymmetric power issues often arise when connecting different branches of science with locally based knowledge systems. It emphasizes complementarity and validation of knowledge within, rather than across, knowledge systems, as well as joint assessments of knowledge contributions (Tengö et al. 2014).

Weaving knowledge from multiple knowledge systems involves five iterative steps: (1) *mobilize*, developing knowledge-based products through a process of innovation and/or engaging with past knowledge and experience; (2) *translate*, adapting knowledge products or outcomes into forms appropriate to enable mutual comprehension in the face of differences between actors; (3) *negotiate*, interacting among different knowledge systems to develop mutually respectful and useful representations of knowledge; (4) *synthesize*, shaping broadly accepted common knowledge bases for a particular purpose; and (5) *apply*, using common knowledge bases to make decisions and/or take actions and to reinforce and feedback into the knowledge systems (Tengö et al. 2017).

This approach has gained traction in IPBES, which, among its core principles, recognizes and respects the contribution of indigenous and local knowledge systems to the conservation and sustainable use of ecosystems. It is linked to the SwedBio program for ecosystem services, a knowledge interface at the Stockholm Resilience Centre. The aim of SwedBio is to facilitate connections across knowledge systems and cultures—such as local, indigenous policymakers and scientific knowledge—and enable dialogue and exchange between practitioners, policymakers, and scientists for development and implementation of policies and methods at multiple scales. The intention is to contribute to poverty alleviation, equity, sustainable livelihoods, and systems rich in biodiversity and ecosystem services.

SwedBio has organized a number of multiactor dialogues, which have contributed to important international policy and practice processes, with relevance for developing countries. For example, the Quito dialogues, convened by several governments together with the CBD Secretariat, helped clarify understanding for action in the challenge of mainstreaming biodiversity. Such SwedBio events have illustrated the power of coproduction of knowledge and transdisciplinary collaboration and has inspired efforts and shaped progress in the CBD.

EAT and the EAT Stockholm Food Forum

Food production is a major driver of global environmental change with significant impact on the biosphere's climate, ecosystems, and freshwater. Food consumption is one of the primary causes of early deaths, with increasing rates of noncommunicable diseases, 800 million people suffering from undernutrition, and over 2 billion people being overweight (Gordon et al. 2017). This dramatic situation led to the creation of the EAT initiative in 2011.

EAT is a science-based global platform for food system transformation that gathers leaders in food, health, and sustainability from science, business, and policy to break the current silo approaches to the food, health, and sustainability agenda. EAT (1) fosters knowledge generation to improve the scientific understanding of the integrated health, food, and sustainability agenda; (2) establishes platforms for engagement of leading change makers in the field; and (3) creates opportunities for actions among key stakeholders.

An example of knowledge generation is the result of the EAT-Lancet Commission that gathered around thirty-five scientists from across the world to define the scientific targets of a healthy diet from a sustainable planet, and the actions needed to assure that 9 billion people have access to healthy food, produced within the environmental limits of the planet (Willet et al. 2019). An example of platforms of engagement is the annual EAT Stockholm Food Forum that attracts 700 world leaders to discuss the latest insights on how to develop a better food system that contributes to sustainability. In terms of action, EAT has, together with the World Business Council for Sustainable Development, set up the FreSH platform, which involves large food companies in science-based dialogues to help spur actions among the companies.

SeaBOS and the Keystone Actor Dialogues

Overexploitation of fish resources is a crucial problem with frustratingly limited scope and scale of progress. We initiated a scientific analysis of globally operating seafood companies engaged in seafood production—actors directly involved with ecosystems. We asked how large the largest companies were, how much they produced, of what species, and from where. We found that thirteen companies were responsible for 11–16 percent of all wild seafood caught, and that together they dominated the industry in white fish, pelagic fish, tuna, salmon farming, and feeds. Through their actions they shape marine food webs and ecosystems, connect ecosystems globally, control globally relevant segments of production through subsidiaries, and they also play a major role in ocean governance through participation in multiple institutions and governing bodies (Österblom et al. 2017).

Following our finding of "Keystone Actors," hypothesized to have a disproportionate ability to enable change in the biosphere, we have engaged in bilateral dialogues with the companies and their CEOs. After two years of such engagement, eight companies were convinced of the value of participating in a global keystone dialogue (November 2016). To our surprise, the companies agreed to form a global coalition for ocean stewardship, based on science, which they named Seafood Business for Ocean Stewardship (SeaBOS). Ocean stewardship was defined as an adaptive and learning-based, collaborative process of responsibility and ethics, aimed to shepherd and safeguard the resilience and sustainability of ocean ecosystems for human well-being.

A second global dialogue took place six months later (May 2017), this time with ten of the companies: they agreed on priorities and a way forward to translate their high-level commitments to operational activities. A third dialogue, with all ten CEOs present, took place in September 2018. The companies reported on real progress made and also committed to long-term funding, an article of association, and leadership of a secretariat, which we as scientists had previously coordinated and facilitated, using independent funding. This large-scale experiment is still in an early phase of development, but companies appear committed to change toward ocean stewardship, and their commitment is already affecting seafood supply chains. It is already inspiring other sectors to engage in strategic leadership for biosphere stewardship.

Global Resilience Partnership and GRAID

The Global Resilience Partnership (GRP) is an independent partnership of public and private organizations joining forces and working toward a sustainable and prosperous future for all, focusing on the most vulnerable people and places in the world. The members of the partnership consider resilience as a prerequisite for advancing the 2030 agenda, including poverty alleviation and development within a safe operating space.

GRP is founded on knowledge excellence and a commitment to finding new ways of dealing with intractable issues of fundamental importance to development. GRP has developed a set of principles that guide its approach to resilience: embrace complexity, recognize constant change, enable inclusive decision making, enhance ecosystems integrity, promote flexibility and learning, and leverage innovation and opportunity. The approach focuses on diagnosing problems, motivating collaboration, developing solutions, and learning and sharing.

The GRP secretariat acts as a support to the entire partnership, including several big actors in the international development sector. GRP has recently moved its secretariat to the Stockholm Resilience Centre (SRC), at Stockholm University, where the GRAID program of the SRC works to support the GRP as its main knowledge partner, sharing its findings and applications in collaboration with academic partners around the world, such as Stellenbosch University in South Africa. GRAID's mission is to develop new thinking, increase awareness, and use resilience insights as an integral part of sustainable development and the international development agenda.

As an example of the collaboration, the GRAID program has developed a new resilience assessment process for planning and action: the Wayfinder. The purpose of Wayfinder is to help development practitioners, project teams, policymakers, and other changemakers work together to strengthen and refine their understanding about the system in focus, the sustainability challenges they face, and the development strategies for creating adaptive and transformative change. A massive open online course (MOOC) on resilience and development was also released in the spring of 2018, together with a rich set of other "flagships" intended to elaborate the way resilience thinking can contribute to new ways of thinking in the development and humanitarian sector.

Finance and Biosphere Stewardship

Humanity now faces the challenge of governing multiple aspects of large-scale global environmental change, such as climate change and its effects, rapid land-use change, biodiversity loss, and many more. While governance can never be neglected as a means to guide human behavior, the urgency of many of the challenges ahead requires identification of additional leverage for effecting rapid, large-scale transformation to safeguard key planetary processes (Abson et al. 2017). Consumers, companies, and even whole value chains have all been the focus of attention as such possible leverage points, but until recently the financial system was not awarded much attention. This has now shifted, and increasing effort within both academia and the financial sector is devoted to understanding how this sector can also contribute meaningfully to change toward a more sustainable future, particularly related to climate (Battiston et al. 2017; Galaz et al. 2018).

While these changes are all welcome, two issues remain urgent for the financial sector to contribute to biosphere stewardship. First, there has been very limited recognition of nonlinear dynamics of many Earth system processes in the strategies and risk scenarios developed to guide sustainable investment. Here, transdisciplinary collaboration and coproduction toward biosphere stewardship is rapidly emerging, including financial actors like insurance companies, banks, pension funds, and hedge funds. Second, to fully understand how financial flows, investment strategies, and capital allocation impact natural capital will require radically increased transparency of financial flows. Such transparency would allow tracing the impact of investments in specific geographical locations and linking them to both measurable improvements as well as holding them accountable for negligible improvements or actual degradation. Achieving such transparency is no small feat but puts the spotlight on key nodes of the global financial machinery and highlights their agency to become active biosphere stewards should they choose to use it.

Conclusions

The seven examples of science for change through transdisciplinary collaboration reflect that it is indeed possible to cooperate with, redirect, and even shift human actions and actors toward biosphere stewardship. They range from engagements with local stakeholders on the ground to those with major players of the globally interconnected world. They illustrate that human actions do not just endlessly degrade the prospect for a positive future, but that there is hope that human actions themselves, and the values, meanings, and incentives behind them, can indeed be transformed toward sustainable pathways and prosperous futures in concert with a healthy and resilient biosphere. This is encouraging and may reflect a broader shift in society moving from an unconscious and mentally disconnected approach to natural capital and environmental issues to increased self-awareness of being part of, embedded within, and dependent upon the planet and its amazing biosphere.

Key References

Abson, David J., Joern Fischer, Julia Leventon, Jens Newig, Thomas Schomerus, Ulli Vilsmaier, Henrik von Wehrden, Paivi Abernethy, Christopher D. Ives, Nicolas W. Jager, and David J. Lang. 2017. "Leverage points for sustainability transformation." *Ambio* 46:30–39.

Battiston, Stefano, Antoine Mandel, Irene Monasterolo, Franziska Schütze, and Gabiele Visentin. 2017. "A climate stress-test of the financial system." *Nature Climate Change* 7: 283–88.

Bennett, Elena M., Martin Solan, Reinette Biggs, Timon McPhearson, Albert V. Norström, Per Olsson, Laura Pereira, Garry D. Peterson, Ciara Raudsepp-Hearne, Frank Biermann et al. 2016. "Bright spots: Seeds of a good Anthropocene." *Frontiers in Ecology and the Environment* 14:441–48.

Folke, Carl, Reinette Biggs, Albert V. Norström, Belinda Reyers, and Johan Rockström. 2016. "Social-ecological resilience and biosphere-based sustainability science." *Ecology and Society* 21 (3):41.

Galaz, Victor, Beatrice E. Crona, Alice Dauriach, Bert Scholtens, and Will Steffen. 2018. "Finance and the Earth system: Exploring the links between financial actors and nonlinear changes in the climate system." *Global Environmental Change* 53:296–302.

Gordon, Line J., Victoria Bignet, Beatrice E. Crona, Patrik Henriksson, Tracy van Holt, Malin Jonell, Therese Lindahl, Max Troell, Stephan Barthel, Lisa Deutsch et al. 2017. "Rewiring food systems to enhance human health and biosphere stewardship." *Environmental Research Letters* 12:100201.

Marshall, Nadine A., Si P. Park, W. Neil Adger, Katrina Brown, and S. Mark Howden. 2012. "Transformational capacity and the influence of place and identity." *Environmental Research Letters* 7:034022.

Österblom, Henrik, Carl Folke, Jean-Baptiste Jouffray, and Johan Rockström. 2017. "Emergence of a global science-business initiative for ocean stewardship." *Proceedings of the National Academy of Sciences, USA* 114:9038–43.

Reyers, Belinda, Carl Folke, Michelle-Lee Moore, Reinette (Oonsie) Biggs, and Victor Galaz. 2018. "Social-ecological systems resilience for navigating the dynamics of the Anthropocene." *Annual Review of Environment and Resources.* doi.org/10.1146/annurev-en viron-110615-085349.

Schultz, Lisen, Carl Folke, Henrik Österblom, and Per Olsson. 2015. "Adaptive governance, ecosystem management and natural capital." *Proceedings of the National Academy of Sciences, USA* 112:7369–74.

Schultz, Lisen, Simon West, Alba Juárez Bourke, Lia d'Armengol, Pau Torrents, Hildur Hardardottir, Annie Jansson, and Alba Mohedano Roldán. 2018. "Learning to live with so-cial-ecological complexity: An interpretive analysis of learning in 11 UNESCO Biosphere Reserves." *Global Environmental Change* 50:75–87.

Steffen, Will, Johan Rockström, Katherine Richardson, Timothy M. Lenton, Carl Folke, Diana Liverman, Colin P. Summerhayes et al. 2018. "Trajectories of the Earth System in the Anthropocene." *Proceedings of the National Academy of Sciences, USA* 115:8252–59.

Tengö, Maria, Eduardo S. Brondizio, Thomas Elmqvist, Pernilla Malmer, and Marja Spierenburg. 2014. "Connecting diverse knowledge systems for enhanced ecosystem governance: The multiple evidence base approach." *Ambio* 43:579–91.

Tengö, Maria, Rosemary Hill, Pernilla Malmer, Christopher M. Raymond, Marja Spierenburg, Finn Danielsen, Thomas Elmqvist, and Carl Folke. 2017. "Weaving knowledge systems in IPBES, CBD and beyond: Lessons learned for sustainability." *Current Opinion in Environmental Sustainability* 26–27:17–25.

Willet, Walter, Johan Rockström, Brent Loken, Marco Springman, Timothy Lang, Sonja Vermeulen, Tara Garnett, David Tilman, Amanda Wood, Fabrice DeClerck et al. 2019. "Our food in the Anthropocene: The EAT-Lancet Commission on healthy diets from sus-tainable food systems." *Lancet: The Lancet Commissions* 393:447–92.

CHAPTER 5

The "Five Ps": Policy Instrument Choice for Inclusive Green Growth

James Salzman

There are many different policy instruments to conserve and provide natural capital in use around the world. Despite the variation among instruments, they are all based on five fundamental policy approaches. These can easily be recognized as the Five Ps: Prescriptive regulation, Property rights, Payments, Penalties, and Persuasion. Once one has mastered the Five Ps, it becomes straightforward to identify the operation, strengths, and weaknesses of specific policy mechanisms.

Difficult choices lie at the heart of conserving and providing natural capital. Should we protect a local population of endangered plants? Should we limit the catch in a fishery that seems in danger of collapsing? And if we take these actions, *how much* should we reduce the threatening activity? Answering these questions is fascinating, but no easy matter. It requires consideration of scientific, economic, legal, and political issues, not to mention the trade-offs that inevitably arise.

And even if we can agree on the goal, a fundamental choice still remains: we need to decide how best to *achieve* this goal. Even if we agree on our starting point and end point, we still need to determine which path should take us there. Reliance on regulatory mandates? Market instruments? Pilot projects or information generation?

While law and policy may appear dauntingly complex, and on occasion truly are, it turns out that understanding instrument choice can be straightforward. Perhaps surprisingly, there are only five basic policy instruments in play, and these can be effectively taught through a simple framework known as "The Five Ps."

Just as a complex sonata can be reduced to a small number of white and black piano keys, so can students' mastery of the Five Ps allow them to identify the potential range of policy instruments available. Despite their application across a dizzying range of situations, the basic policy tools remain the same.

The **Five Ps** are **Prescriptive** Regulation, **Property** Rights, **Penalties**, **Payments**, and **Persuasion**. There will rarely be one best tool for a particular situation, and much of the challenge in instrument choice lies in identifying each instrument's particular advantages and disadvantages.

The Five Ps analysis can be applied to any environmental problem. In this chapter, we examine it in the context of conserving and providing natural capital. In the sections below, each policy approach is set out and applied to the classic example of the *tragedy of the commons* (Hardin 1968). In this case, too many sheep are grazing on an open-access commons. This is leading to degradation of the grassland, resulting in the loss of ecosystem services such as forage, soil retention, and flood control. Unless something is done, the grass will soon be overgrazed and these services will be completely lost. Individual incentives encourage rapid depletion of the resource to the detriment of all—hence the tragedy.

The goal of managing the commons at a sustainable level is clear. The question is which policy instrument should be used to achieve this goal.

Prescriptive Regulation

Prescriptive regulations mandate what parties can and cannot do. This is both the most direct and the most common form of environmental law. We see prescriptive regulations at all levels of environmental governance—from hunting permits at the local level, and effluent limits under pollution laws at the national level, to restrictions on foreign commerce in endangered species at the international level.

In the context of overgrazing the commons, for example, the government might limit the number of sheep that may graze, or restrict grazing to a particular season or period of time.

Also referred to as *command-and-control regulation*, prescriptive regulation can be very effective in mandating uniform compliance across all actors, preventing problems of hold-outs, free riders, and collective action. We can see this policy instrument in chapter 7's discussion of regulatory-driven mitigation, where prescriptive regulation limits development of wetlands or streams unless mitigation is provided.

There is considerable debate over the efficiency of prescriptive regulations. Economists, for example, often criticize them as inefficient and unwieldy. They argue that this approach provides little incentive for innovation because once the regulated party has satisfied the necessary requirement, the law creates no incentive to reduce harmful activities further.

Two unstated assumptions behind prescriptive regulation merit mention. The first is that the regulator will set the standard at the proper level. This may not happen, either because of inadequate information or agency capture (a classic problem in the context of natural resources such as fish and timber, where industry pressure has led to overfishing and large-scale clearcutting). The second assumption is that the regulator will be able to monitor compliance with the standard. Both impose administrative costs, which, as result, can sometimes be a good deal higher for prescriptive regulation than for other policy instruments.

Property Rights

A classic solution to the tragedy of the commons is to privatize the resource by creating property rights. In the grazing example, instead of an open-access commons, assume the field has now been divided into square parcels of land and allocated to individual shepherds, including you. You now have the right to exclude everyone else's sheep from your parcel. Are you still as eager to overgraze as before?

All of a sudden, your previous incentive to consume the resource as fast as possible (before everyone else does) is no longer relevant. Instead, your interests are best served by carefully tending your part of the commons so it remains productive long into the future—so it is *sustainably managed*. You may well charge other shepherds to use your parcel, or even let them on for free, but you would do so only to the extent that the resource base remains intact and productive—that is, so long as the resource is not overgrazed. In financial terms, to maximize profits you will safeguard your asset over the longer term. The same should be true whether the property rights are vested in individuals or in communities (as is the case in many indigenous cultures).

Compared to prescriptive regulation, this approach should have lower administrative costs. The government simply creates the property rights, allocates them initially, and steps back, leaving future allocations to the market.

We see this policy instrument in chapter 10 in the example of water rights in Australia's Murray-Darling Basin.

Implicit in a property-rights approach is the importance of technology. To enforce your right to exclude, you need both to know someone is making use of your resource (an issue of monitoring capacity) and to have the ability to exclude others' use.

Despite the increasing interest and application of property-rights approaches to environmental protection, they face some significant obstacles. The first is that many environmental resources are not easily amenable to commodification. When resources have significant public-goods aspects (such as major watersheds or biodiversity), privatization might not lead to the most socially beneficial use of the land. Private-property owners typically value only those uses that provide monetary remuneration. In these cases, the important positive externalities may not be valued. If the government wants to ensure the important public goals of a secure food supply, conservation of rare biodiversity, or buffers against flooding, it may need to step in and restrict the use of the land.

Practically, there also are difficult allocation issues for the initial privatization of environmental resources. Using the commons as an example, assume that the government has divided up the land into fifty separate parcels. Who should be given title? Should the land be auctioned to the highest bidder? This could favor wealthier newcomers and corporate interests. To give more respect to traditional users, perhaps the allocation could be based on historic use or current levels of consumption? Yet this would put newcomers at a disadvantage and favor those who have been the most profligate in the past. If we cannot decide among these competing users, should we just have a random drawing? Any allocation mechanism will tend to favor some groups at the expense of others. Inevitably, who should be favored comes down to a contentious political decision, with winners and losers.

Tradable Permits

Prescriptive regulations can be combined with property rights through the use of tradable permits in environmental markets. Here, property rights are created for *use* of the resource—for the right to graze sheep in a certain area, for example. Trading systems use the market to make prescriptive regulation more efficient. The government decides how much of a harmful activity to permit (just as it would with prescriptive regulations), awards private rights to engage in the activity up to the regulatory cap, and then permits those rights to be traded. The market does not play a role in determining the overall level of environmental protection; that is the role of the regulatory regime.

To make this more concrete, imagine how a trading program would work with grazing on the commons. Policymakers decide that the commons can sustain no more than 400 sheep grazing per year. The government therefore creates 400 permits, entitling the holder to graze one sheep for the calendar year listed on the permit. Unless the shepherd has a separate permit for each sheep grazing on the commons, she is breaking the law. The government then allocates the permits in some fashion (which, as noted above, will have significant distributional consequences) and lets trading commence. Those for whom grazing is most valuable will pay the highest price to buy the permits from those who value it less, ensuring that the commons is dedicated to the most valuable market use. If the cap is set appropriately, marketable permits achieve the same level of protection as command-and-control alternatives, but at a lower cost. This is the ultimate goal of the REDD+ programs described in chapter 11, when carbon sequestration credits can be sold to countries to meet their climate change goals.

Initially allocating permits presents a challenge for trading, just as it does for pure private-property approaches. Moreover, constructing smoothly functioning markets is not simple. There must be a well-defined marketplace and enough buyers and sellers to support an active market. There also must be an effective currency of trade, one that is fungible and that reflects the desired environmental quality. For example, it would be a stretch to consider allowing coastal developers in Florida to "trade" the wetland ecosystem services they eliminate (such as flood control or nutrient filtering) for phosphorous emissions reductions in Oregon.

Because the market decides where the allowances go after their initial allocation, there is a further challenge that harmful activities can be concentrated, creating local "hotspots" of pollution. This can become an environmental justice concern, for example, when harms are concentrated in low-income communities.

Financial Penalties

Short of banning an activity, another effective way to limit behavior that degrades natural capital is to make it more expensive, whether through charges, taxes, or liability. By increasing the costs of harmful activities, such penalties force the parties to bear the costs of their activities. To use economics language, the polluter internalizes the negative externalities of her behavior. In our commons example, shepherds might be charged a fee per sheep for the right to graze each day. The fee could be shifted up or down, depending on the desired level of grazing

In theory, financial penalties offer an attractive policy instrument but there are two practical obstacles. The first lies in getting the price right. Markets are efficient when the prices for goods accurately reflect their full environmental and social cost. A key aspect in internalizing externalities, then, is valuation. If one agrees that externalities should be internalized—that parties should pay for the harm caused—the obvious question is "how much?" In practice, though, it may be sufficient to ignore this and focus simply on the amount necessary to bring about the desired behavior change. This approach would focus not on the value of the harm caused but on the marginal opportunity cost.

The second challenge is political. Increasing taxes is never easy, and environmental charges seem to be harder still. And levying them at charges high enough to influence behavior significantly is easier said than done. In many cases, the charges have been intended more for revenue-raising than for serious behavior modification.

Financial Payments

As noted above, government can discourage certain activities through penalties but, equally, it can use subsidies to encourage beneficial activities. Just as government can use penalties to capture negative externalities and make bad activities more expensive, it can use payments to capture positive externalities and make good activities less expensive. In our commons, shepherds might be paid $100 *not* to graze one of their sheep for a year. The shepherd is effectively being paid for not exercising her right to graze. In one example of this approach, government can provide agricultural subsidies to farmers for setting aside cropland to prevent erosion or provide wildlife habitat. This is the approach behind the popular strategy of payments for ecosystem services described for domestic payments in chapter 6, water funds in chapter 9, and multilateral payments in chapter 11.

Not all payment schemes benefit the environment, however. Quite the opposite, since many government subsidies actually encourage harmful activities. Independent analyses have identified billions of dollars in subsidies whose elimination would both help the environment and reduce the federal budget deficit (such as subsidies for building logging roads on public lands). In certain respects, perverse subsidies cost us twice—first, when we pay the initial tax to raise the funds needed for the subsidy, and second, when we suffer the environmental damage and loss of natural capital encouraged by the subsidy.

Persuasion

If prescriptive regulation and market instruments represent "hard" regulatory approaches, then a softer approach may be found in laws requiring information production and dissemination. The theory behind such approaches is that the government can change people's behavior by forcing them to think about the harm they are causing and by publicizing that harm. In the context of the commons, the government might require shepherds to record and publish the number of sheep they graze, the amount of forage the sheep eat, or the days before the commons can no longer support grazing. Landowners might be required to create management plans demonstrating how they will increase their riparian fencing over a five-year period. The government also may try to educate the shepherds with brochures or presentations on the causes and dangers of overgrazing, or it may sponsor a pilot project that demonstrates more effective ways to manage the commons.

Information-based approaches are often used when there is inadequate political support to impose market or regulatory instruments, or when such instruments are ill-suited to the problem. Whether because of "naming and shaming," measuring harms for the first time, or heightened consciousness, this persuasive instrument has led to significant reduction of harmful activities in a range of settings.

Putting the Toolkit to Work

While the examples above have used the case of natural capital loss from grazing on the commons, one can apply this toolkit of regulatory instruments to virtually any environmental problem. Taking the service of water purification as an example, consider the situation where the drinking water of a community is polluted by the manure of cows in fields and streambeds. The city could build a treatment plant to purify their drinking water, but officials have determined that it will be much less expensive if strips of vegetation are erected alongside the streambanks leading into the reservoir. These riparian buffers will provide the ecosystem service of nutrient retention, with the plants taking up the nitrates and phosphates before they flow into the streams. If you were head of the local government, what would your proposed riparian buffer strategy look like?

Prescriptive regulation could require landowners to place fencing along streambeds to protect riparian vegetation and keep cows out. Conversely, regulation could prohibit cattle from entering streambeds.

Financial penalties seem a good potential fit as well. You could levy fees based on the length of a streambed that did not have riparian fencing. Conversely, you could rely on payments, providing tax credits or subsidies per meter of riparian fencing in place. While an unwieldy approach, one could use a property rights strategy by establishing a trading market for cattle allowed to graze in streams. The cap could be set very low, allowing landowners to choose between putting in fencing or purchasing stream grazing allowances.

Finally, persuasion through information disclosure might work well. You could, for example, require landowners to collect and publish data on their riparian fencing or number of cattle that enter the streambeds.

The Five Ps could also apply to conserving an endangered species, perhaps a rare bird. Prescriptive regulation might ban actions that kill or harm the birds, or degrade their habitat. Property rights could be used to create a trading program where breeding pairs became the currency of exchange—landowners who modified their habitat so it was less attractive to the birds could mitigate their actions by purchasing credit for breeding pairs that had been established by entrepreneurs in other areas. This "species banking" could create an incentive for entrepreneurs to convert farmland, for example, into endangered species habitat. Financial penalties could be imposed on landowners who make habitat less attractive to local endangered species. Conversely, payments could be made to landowners who improve habitat to make it more attractive to breeding pairs. And persuasion could be used by highlighting the natural heritage of local biodiversity.

To be sure, this framework cannot perfectly capture the dizzying range of all environmental policy. The vast majority of instruments, however, do fit easily within the Five Ps framework. As a result, policymakers can quickly assess the toolkit available for any specific threat to natural capital, considering the relative strengths, and comparative weaknesses of instrument choices.

Key Reference

Hardin, Garrett. 1968. "The tragedy of the commons." *Science* 162 (3859):1243–48.

PART II

Policy and Finance Mechanisms for Natural Capital, Ecosystem Services, and Livelihoods

CHAPTER 6

Government Payments

Lisa Mandle

Many ecosystem services originate on privately owned land but benefit the wider public. Without incentives to maintain these benefits, landowners may prioritize management for private returns at the expense of public benefits. Government institutions are often well positioned to aggregate funds on behalf of many beneficiaries and to provide compensation to service providers. This chapter presents three examples of this mechanism: the Conservation Reserve Program in the United States, the protection of New York City's drinking water supply, and the Working for Water program in South Africa. In these case studies, national or city governments collect funds through taxes, bonds, or utility rate payer fees. These funds are then distributed to individuals or institutions in return for implementing management practices that enhance ecosystem service provision. In addition to improving public ecosystem service benefits, these programs also seek to improve rural livelihoods. Government payment programs for ecosystem services, like the examples presented here, are one of the longest-standing and most widespread mechanisms for securing and enhancing natural capital.
....

The connection between conservation of watershed lands and downstream water quality and quantity have long been recognized. In France, the protection of forests and water have been linked since the thirteenth century (Andréassian 2004). Thus it is no surprise that government payment programs for ecosystem services are vital and longstanding, especially when it comes to water-related ecosystem services.

In this chapter, we describe three examples of how government payments have been used to secure and enhance environmental benefits. The first two examples come from the United States: the cases of the federal government's

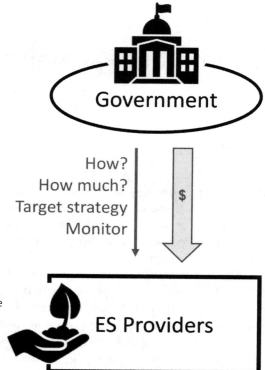

Figure 6.1. Under the government subsidy mechanism, the government compensates ecosystem service providers with funds from its general budget. The government determines where and how to pay for securing or enhancing ecosystem services and is responsible for monitoring outcomes.

Conservation Reserve Program (created in the 1980s), and New York City's water supply (created in the 1990s). The third example, the case of Working for Water (also created in the 1990s), comes from South Africa. In the case of New York City's water supply, the service providers are compensated by the New York City beneficiaries. The Conservation Reserve Program and Working for Water both operate primarily through government subsidies, in which ecosystem service providers are paid by the general tax base, rather than by specific beneficiaries of the service (fig. 6.1).

Case 1. The United States Conservation Reserve Program

The Problem

In the 1970s in the United States, agricultural policy and domestic demand led to steep increases in the price of agricultural commodities. In order to increase supply, American farmers were encouraged to plant "fencerow to fencerow." As a result, forests, shrublands, and grasslands, along with other extensive marginal, low-productivity lands were converted to crop production. This had detrimental environmental effects, including increases in soil erosion, decreases in water quality from runoff of agricultural pollutants, and loss of habitat for wildlife. The increases in production, combined with international geopolitical factors, also depressed crop prices for farmers. The Conservation Reserve Program (CRP) was established in 1985, following other US government programs going back to the 1950s that encouraged agricultural land retirement The CRP was created to reduce the environmental impacts of agricultural production, while also supporting farmers during a period of low crop prices. It provides subsidies to farmers to convert environmentally sensitive croplands to perennial vegetation in order to improve environmental conditions. Thus the program has both economic development and environmental protection goals.

The Ecosystem Services

The CRP has three main environmental objectives: to improve water quality, reduce soil erosion, and improve wildlife habitat. Fertilizers and pesticides applied to crops can run off into waterways, diminishing water quality. Tilling leaves topsoil exposed to wind and rain, exacerbating erosion and reducing soil fertility, air quality, and water quality. Conversion of natural vegetation to farmland reduces habitat available for wildlife species. The CRP's objectives are linked to a variety of ecosystem services with economic and health benefits, including improved drinking water quality; improved marine and freshwater fisheries resulting from improved water quality; recreational opportunities from improved water quality and increased wildlife viewing and hunting opportunities; and improved air quality from reduced wind-borne dust.

Ecosystem Service Beneficiaries

Many people stand to benefit from the ecosystem services provided by the CRP. The specific beneficiaries depend on the location of CRP activities relative to where and how people rely on clean water, clean air, and benefits from wildlife. Unlike the case of New York City's water supply, here the beneficiaries do not pay directly for the ecosystem service benefits, which are instead supported by the federal tax base (see *Mechanism for Transfer of Value* for more details.)

Ecosystem Service Suppliers

Farmers whose land-use decisions and management practices affect water quality, soil erosion, and wildlife habitat are the ecosystem service suppliers. Eligible farmers who choose to enroll in the CRP are compensated. Public lands are not included.

Terms of the Exchange: Quid Pro Quo

Under the CRP, farmers who enroll eligible land agree to remove that land from crop production and plant perennial vegetation to improve environmental quality in exchange for an annual rental fee over the ten- to fifteen-year duration of the contract. To be eligible, land must have been in production for four out of six of the previous years, or be marginal cropland or pastureland near a water source. The CRP also provides up to 50 percent cost share (paying for up to 50 percent of the total costs) and up to 20 percent of the annual rental fee for additional conservation activities like windbreaks, filter strips, and riparian buffers.

The CRP provides approximately US$1.9 billion per year in funding. At its peak in 2007, 15 million hectares were enrolled. As of 2017, enrollment had declined to less than 10 million hectares—the lowest level since 1988—reflecting the cap the government set in 2014 aimed at reducing the CRP budget (fig. 6.2).

Mechanism for Transfer of Value

The CRP is funded under the US Farm Bill and administered by the Farm Service Agency (FSA) within the US Department of Agriculture (USDA). Farmers enter into contracts with the Commodity Credit Corporation, also under the USDA. Contracts are awarded based on a reverse auction. Landowners submit bids for a requested rental rate for a specific conservation activity (e.g., planting grasses, trees, or riparian buffers) on a specific parcel. Bids are accepted during an approximately annual enrollment window. Bids submitted during this time are then ranked against each other based on an Environmental Benefits

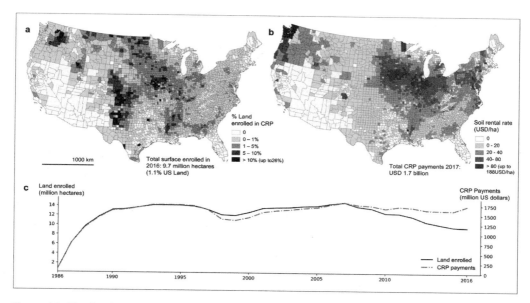

Figure 6.2. The distribution of land enrolled in the Conservation Reserve Program (CRP) across the United States (*a*); average soil rental rates (US$/ha) for CRP land (*b*); and the total area of land enrolled in CRP and CRP payments from 1986 to 2016 (*c*).

Index (EBI). EBI scores are calculated based on the parcel and its associated conservation activity's value for water quality, erosion control, wildlife, as well as the parcel's rental rate. Bids are accepted in order of their EBI score, up to a maximum target acreage. In addition, CRP acreage cannot exceed 25 percent of the cropland per county.

Because of the acreage limits and the limited enrollment time window, farmers are competing against each other to win a contract and enroll in CRP. Farmers can increase the likelihood of their bid being accepted into CRP by increasing their EBI score in two ways: by lowering their requested rental rate, and/or by offering to implement an activity on the parcel with greater conservation value. The FSA sets the maximum rental rates for the contracts. The rates vary by parcel, based on soil productivity and crop value, and these rates are provided to landowners to be used in preparing their bids.

Monitoring and Verification

The FSA is supposed to monitor CRP contracts to ensure they comply with their required conservation practices and acreage enrolled. However, according to the most recent assessment by the United States Office of Management and Budget in 2005, monitoring has been problematic. There is often not an effective response when land is not being managed according to the contract.

Effectiveness

The USDA estimates that the CRP has prevented over 8 billion metric tons of soil from eroding, as well as reduced nitrogen runoff by 95 percent and phosphorous runoff by 85 percent, relative to annually tilled cropland (USDA 2015). One million hectares of wetlands have been restored and 275,000 kilometers of streams have been protected with riparian buffers. In addition, the CRP has promoted the sequestration of an average of 44 million metric tons of greenhouse gases annually. The benefits to wildlife have also been substantial, especially for grassland birds and waterfowl.

What these biophysical changes mean in terms of economic values or other benefits to people is less clear. A 2007 study estimated that the program provided US$1.3 billion in annual benefits nationally, which represents 75–80 percent of the program's costs (Hansen 2007). However, the study valued only a subset of the program's total benefits, so this represents a very conservative estimate. Earlier assessments concluded that CRP benefits outweigh program costs.

The CRP could more effectively target the places where reduced erosion and improved water quality matter most to people and increase the program's economic benefit by adjusting the method used to prioritize land for enrollment based on its environmental benefits. Scores from the CRP's Environmental Benefits Index (EBI) are based largely on the biophysical aspects of ecosystem services, or the *potential* for a particular parcel to improve air quality and water quality regulation services. EBI provides some weighting for land that occurs in high priority water quality and air quality zones. However, a 2008 CRP program assessment suggests that this is not enough, and providing additional weight to land in priority areas could improve the program's contribution to solving water quality and air quality problems at national and regional scales (Soil and Water Conservation Society and Environmental Defense Fund 2008).

Adjustments to the design of the CRP's reverse auction enrollment mechanism could also increase the program's effectiveness (Hellerstein 2017). Over time, the bids accepted into CRP have converged toward the maximum rental rate for parcels. Because there are recurring enrollment periods approximately annually, and because most bids are accepted, farmers whose bids are not accepted one year can apply again the next year using the information they gain to adjust their bids accordingly. Farmers can therefore be aggressive in their bidding and capture more rent for providing the same benefit than they might otherwise. Increasing competition among farmers and inducing bids closer to the real opportunity cost through changes to EBI or the auction mechanism could make CRP more cost effective.

Finally, environmental groups have raised concerns about the durability of CRP benefits. At the end of a ten- to fifteen-year CRP contract, landowners can choose whether or not to reenroll their land. When crop prices are high relative to the maximum rental rates, land is often not reenrolled and is instead converted back to agricultural production. For example, between 2007 and 2014, a period of rising crop prices, 6.5 million hectares of CRP lands were not reenrolled when their contracts expired (Schechinger and Cox 2017). A number of groups have suggested that acquiring long-term or permanent easements would secure environmental benefits more cost effectively, despite the higher cost per acre (Schechinger and Cox 2017).

Key Lessons Learned

The CRP has successfully enrolled millions of hectares, leading to important reductions in sediment and pollutants in waterways, improvements to air quality, and enhancement of wildlife habitat. The subsidy support provided to farmers, especially in times of low crop prices, has been key to attracting political support and funding. The CRP's reverse auction system for enrolling land is the program's main policy innovation. While there may be opportunities for further increasing its effectiveness, the reverse auction mechanism has secured environmental benefits more efficiently than offering a simple per-acre price.

Sustaining the benefits of the CRP is a challenge, especially in the face of increasing crop prices. Because CRP contracts come up for renewal every ten to fifteen years, if crop prices are high, farmers may choose not to reenroll and instead put lands back into agricultural production, making the environmental gains transient. In addition, funding for the program under the Farm Bill must be renewed approximately every five years, and securing adequate funding in an ever changing political climate can be difficult.

Case 2. New York City's Water Supply

The Problem

New York City's water system provides 4.5 million cubic meters of water per day to over 9 million people in and around the city. All of the water comes from surface sources: three watersheds located 80–200 kilometers outside the city—the Croton, Catskill, and Delaware watersheds (fig. 6.3). Giant pipes transport the water to reservoirs outside the city, where it is chlorinated and then distributed to city water mains. The result is urban drinking water that has long been recognized as among the very best in the United States.

In response to an outbreak of water-borne disease (*Cryptosporidium*) in Milwaukee, Wisconsin, which led to more than 100 deaths and 400,000 illnesses, the United States Environmental Protection Agency (EPA) implemented the Surface Water Treatment Rule in 1989. The new rule required large public water systems that rely on surface sources to filter their water regardless of quality. Based on this, New York City built a treatment plant to filter water from the nearest watershed—the Croton, which was most impacted by suburban development and supplies 10 percent of the city's water. Constructing this filtration plant came at an eventual cost of over US$3 billion, well over an early estimate of US$800 million (Dunlap 2015).

The EPA rule also had a waiver provision, however, that allowed the city to avoid building a treatment plant if it could demonstrate that it had taken other steps necessary to maintain safe drinking water. This left New York City with two options for the Catskill/Delaware watersheds, which supply the other 90 percent of its drinking water: build an additional filtration system at an estimated cost of US$6 billion or more (plus annual operating expenses of US$300 million; Chichilnisky and Heal 1998), or petition for an exemption by showing that it could maintain safe, high-quality drinking water through effective management of its unfiltered source watersheds. Based on a comparison of costs between filtration and watershed protection (estimated at US$1.5 billion), the city chose to work with and invest in upstream communities in order to deliver safe, clean drinking water without filtration. This agreement was formalized in 1997 with the New York City Watershed Memorandum of Agreement, and it remains in place today.

The Ecosystem Service

Watershed protection contributes to regulation of water quality in two ways. First, maintaining natural vegetation and following best management practices

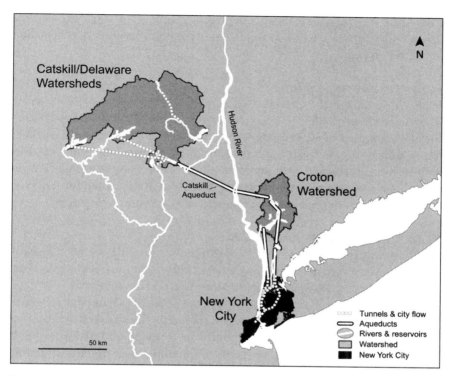

Figure 6.3. New York City's drinking water supply originates in the Croton, Catskill, and Delaware watersheds, located between 80 and 200 km outside the city. Since 1997, the city has worked with and invested in communities in the Catskill and Delaware watersheds to protect water quality regulation services and deliver safe, clean drinking water without filtration.

reduce the amount of pollutants entering the system. Pollutants include those from discrete point sources, such as wastewater treatment plants and sewage systems, and from diffuse nonpoint sources, such as dairies and apple orchards. Second, vegetation acts as a natural filtration system, capturing water and directing it to shallow subsurface flow, where soil and vegetation can sequester pollutants and keep them from reaching streams and reservoirs.

Ecosystem Service Beneficiaries

The beneficiaries of maintaining clean water are the water consumers (rate payers) in New York City, who otherwise would face increased water and sewer rates if an expensive filtration plant needed to be built. According to estimates from the New York City Independent Budget Office in 2000, the average household would face fees that were 13 percent higher with the construction of a filtration plant.

Ecosystem Service Suppliers

With 70 percent of the Catskill/Delaware watersheds under private ownership, dairy farmers, apple growers, foresters, and other rural landowners are the main suppliers of water quality regulation services. Farmers who enroll in the voluntary Whole Farm Program and follow best management practices defined by the program, in order to minimize impacts on water quality, receive direct compensation for these activities. In addition, communities in the source watersheds receive funding for general economic development.

Terms of the Exchange: Quid Pro Quo

New York City initially provided US$350 million in funding in 1997 for a variety of activities aimed at maintaining water quality of the upstream watersheds. As of 2018, the city has spent approximately US$1.5 billion in total.

Some of this funding contributes directly to improving water quality and is not spent on ecosystem service provision. For example, wastewater treatment plants have been upgraded to reduce point-source pollution, and the use of salt on icy roads during winter has been reduced. Funding also supports economic development and job opportunities in the upstream watersheds. This funding comes in exchange for restrictions on development and land use agreed upon under the Watershed Rules and Regulations. Many of these restrictions protect water quality regulation services. For example, they prohibit new impervious surfaces near waterways, wetlands, and reservoirs in order to protect the water filtration capacity of vegetation in these areas.

In addition, farmers are paid directly through the Whole Farm Program to improve management practices and enhance ecosystem service provision on their land. This includes implementation of best management practices such as installation of vegetated filter strips, which trap sediments and nutrients in agricultural runoff, and fencing to keep livestock out of wetlands and streams.

Finally, some funding is used to buy land outright or purchase easements at a fair market value from willing landowners. This allows New York City to acquire and otherwise protect land around reservoirs and their tributaries, buffer areas that are key sources of water filtration services. Altogether, approximately 30,000 hectares are under improved management and 50,000 hectares of land or easements have been purchased.

Mechanism for Transfer of Value

The transfer of funding from New York City to service providers in upstream watersheds occurs under the New York City Watershed Memorandum of Agreement (MOA). Funding for the MOA comes from a combination of rate payer fees and from bond funds ultimately supported by taxpayers. The MOA—that includes New York City, New York State, the US Environmental Protection Agency, a coalition of thirty watershed towns, and environmental groups—was formalized in 1997. Somewhat unusually, state law gives New York City the power to regulate land use and development activities in its source watersheds in order to maintain the quality of its drinking water. The power to regulate land use over 150 km outside the city limits is very rare in United States law. It was the threat of unilateral regulation that brought upstream communities to the table to work out a bilateral agreement with the city.

The MOA established the Catskill Watershed Corporation to administer many of the community-level programs for protecting water quality and promoting economic development. The Watershed Agricultural Council implements the Whole Farm Program, described previously in *Terms of the Exchange*. Participation in the Whole Farm Program is voluntary for individual farmers. However, upstream communities needed to get the participation of 85 percent of landowners within five years; otherwise, they would face unilateral land-use regulation by New York City. In the end, the voluntary farmer participation rate exceeded 90 percent.

Monitoring and Verification

The quality of water from the Catskill/Delaware watersheds is monitored intensively and remains high. New York City's Department of Environmental Protection monitors compliance with the Watershed Rules and Regulations.

Effectiveness

From a biophysical perspective, watershed management has been successful at maintaining water quality that meets federal and state safety and quality standards. New York City's water supply permit and filtration exemption from the EPA was renewed in 2002, 2007, and 2017. The city's water supply remains the largest unfiltered source of drinking water in the United States. The MOA was renewed for another fifteen years in 2012 with the support of watershed communities. This suggests that the social dimensions of the program are deemed successful to communities, as well.

Key Lessons Learned

As the balance of costs between watershed management and a water filtration plant demonstrates, natural infrastructure can be more cost-effective than built infrastructure at delivering services, even without including the cobenefits from conserving natural infrastructure, such as recreational opportunities and aesthetic values. This example also illustrates how a variety of mechanisms—from direct support to landowners, to general support of economic development of the supplying region, to outright land acquisition—can be combined to achieve the desired provision of water quality services.

It is important to note that the regulatory environment, and specifically the 1989 rule change that required either filtration or active watershed management, set the stage for New York City's investment in upstream watersheds. Without the regulatory mandate, it is not clear at all that New York City would have seriously investigated the payment scheme to watershed landowners and communities, much less raised the funds to pay for service provision. The city's ability under state law to regulate land use in upstream watersheds was another critical factor that pushed rural landowners to work cooperatively with government officials and generate a voluntary agreement.

Even keeping these specific conditions in mind, the Catskills program represents the best known and arguably one of the most successful payment for ecosystem service schemes. Faced with a regulatory mandate to ensure safe drinking water for its residents, New York City determined that simply in terms of pure market finance considerations, investing in green, natural capital through payments for improved land use was a better decision than investing in grey, built capital such as a water treatment plant. It comes with three wins: a financial win for the city and its beneficiaries; a win for upstream suppliers in terms of financial and other support; and a win for the many other dimensions of natural capital and their ecosystem services protected under the safe drinking water umbrella.

Case 3. Working for Water in South Africa

The Problem

Water scarcity is a major challenge for the country of South Africa. With a semiarid climate, water availability is limited and fluctuates dramatically both throughout the year and between years. In addition, water resources are

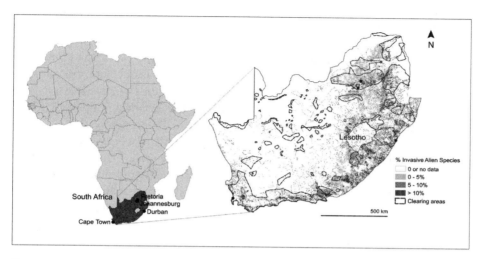

Figure 6.4. Extent of invasive alien species in South Africa (*shown in gray shading*), along with areas cleared by Working for Water (*shown with a dashed border*). Note that these areas show where at least some clearing has occurred, but not all land within these areas has been fully cleared.

distributed unevenly across the country, with 50 percent of the water flowing in rivers and streams coming from just 13 percent of the land. Water requirements exceed availability for much of the country. Without increases in water use efficiency, demand for water is projected to exceed supply in the country as a whole by 2025. Historically, water demands were met through engineered solutions, with construction of 800 large dams to store water and twenty-eight interbasin transfer schemes, including import of water from neighboring countries. However, opportunities for increased water supplies through additional dams and transfers are limited and expensive.

Starting in the 1970s, there was a growing realization in the scientific community that the spread of water-thirsty, invasive alien plants threatened not only the country's native plant biodiversity but also its water resources. Without funding to control the spread, stream flows in the western part of the country were likely to decline between 20 and 50 percent, and water supplies to the city of Cape Town could potentially be reduced by 30 percent (van Wilgen and Wannenburgh 2016). The Working for Water (WfW) program was launched in 1995 as a public works program to support clearing of invasive alien plants in order to create jobs and secure South Africa's water resources (fig. 6.4).

The Ecosystem Service

Invasive alien plants in South Africa, including trees such as pines, eucalyptus, and wattles, as well as riparian woody species, use more water than the native vegetation they replace. Through their roots, they transfer water from underground (where it feeds into streams) to the atmosphere (where it is lost to local people). Catchment-scale studies have shown that removal of these alien species leads to increased stream flows, especially dry-season flows.

Clearing of invasive alien species can be a cost-effective option for increasing water supplies, as several studies undertaken in the early 1990s demonstrated. An analysis of the water supply for the Cape Town metropolitan area showed that controlling invasive species in the existing source catchment would provide water at 13.6 percent of the cost as compared to building a new dam to capture water from an additional catchment (van Wilgen et al. 1997). Clearing vegetation was shown also to be more economically favorable for increasing water supply than desalination or reuse of sewage.

Ecosystem Service Beneficiaries

Water users—including cities, industry, and agriculture—benefit from the increased water availability that comes from clearing invasive alien species. For the most part, beneficiaries do not pay directly, but there are some exceptions (see more in *Mechanism for Transfer of Value*).

Ecosystem Service Suppliers

The suppliers of the ecosystem service of water provision are the government and the people who clear invasive alien plants from the land. Compensation is provided to the people who clear the plants, rather than to the landowners. Much of the land cleared under WfW is public land. Because the people clearing the land are generally from low-income communities, WfW serves the dual purposes of economic development and environmental protection. Indeed, as described below, WfW's primary focus is poverty alleviation.

Terms of the Exchange: Quid Pro Quo

WfW began in October 1995 with ~US$2.5 million and 10 projects in six out of nine provinces (van Wilgen and Wannenburgh 2016). As of 2016, this had expanded to 300 projects across all nine provinces, with a budget of over US$75 million. Since its start, 2.5 million hectares have been cleared at least once, with follow-up clearing 2.7 times on average (fig. 6.4). By 2015, 40,000 people had been employed through WfW. It has spurred the development of a portfolio of similar programs with other environmental objectives, such as Working for Wetlands, aimed at wetland rehabilitation, and Working on Fire, aimed at fire management.

Mechanism for Transfer of Value

Implementation of WfW is based on a contractor system. WfW appoints implementing agents that are responsible for managing projects on the ground. Institutions or organizations such as municipalities, government conservation agencies, irrigation boards, or forestry companies serve as implementing agents. Implementing agents then recruit and manage contractors, who are individuals that have set up their own small businesses and contract with WfW. Contracts are awarded through a noncompetitive bidding system, though there is pressure to change this to a competitive system. Contractors are assigned specific areas to clear and paid at a rate based on the particular species present and their density. Contractors employ the teams of workers who carry out clearing and management of invasive alien plants.

At its start, WfW was funded by the South African Government's Reconstruction and Development Programme, which focused on reconstruction and socioeconomic development in the post-Apartheid era. WfW is currently funded primarily by the Expanded Public Works Programme in the Department of Public Works. The program is administered by the Department of Environmental Affairs. According to the most recent budget estimates available, around 80 percent of WfW's budget comes from the central government's Poverty Relief Fund, generated from general tax revenues (Ferraro 2009). Approximately 10 percent of the budget comes from water resource management fees charged to water users in certain Water Management Areas; this funding is earmarked for removal of aquatic weeds. Some municipalities, state-owned utilities, and private companies have paid the WfW program for clearing specifically within their catchments (Turpie, Marais, and Blignaut 2008). Foreign donors and other government departments (e.g., Tourism, Agriculture) have also contributed funding. WfW has also experimented with attracting cofunding from landowners to clear invasive alien plants from private lands, though this is still a minor component of the WfW program.

Monitoring and Verification

Monitoring and verification addresses both biophysical and social dimensions of the program. Control efforts are monitored in terms of the area of invasive alien plants treated, the number of sites where biological control agents are established, and the number of emerging invasive alien species controlled. On the social side, the focus is the number of full-time equivalent jobs created, with sub-targets for women, young people, and people with disabilities. Because the program is funded by the Expanded Public Works Programme, which is aimed at providing work opportunities for the unemployed, success is evaluated primarily in terms of costs per person per day, with the goal of minimizing costs and maximizing employment opportunities. The ecosystem service outcomes, in terms of changes in water resources, are not monitored, though they have been estimated through modeling (see more in *Effectiveness*).

Effectiveness

Between 1995 and 2017, WfW created over 630,000 work opportunities and more than 230,000 full-time-equivalent job years (Wannenburgh 2018). The program has succeeded in providing temporary employment opportunities to many people. Whether it has improved long-term employment prospects for people once they leave the program, though, is not clear (Coetzer and Louw 2012).

WfW has funded the clearing of invasive plants from large areas. Although the effectiveness of the program in securing water resources or other ecosystem services is not directly monitored, the government estimated that between 50 and 130 million cubic meters per year were made available by clearing through 2003, equivalent to between 0.1 and 0.3 percent of mean annual runoff in the country (Gorgens and van Wilgen 2004).

However, the spread of invasive alien species is still outpacing control efforts in many areas. Because of the program's job-creation goals, clearing efforts are not necessarily directed at the highest priority areas from a water resources perspective. Scientists have mapped priority areas for invasive alien plant removal; however, projects continue in low-priority areas because of political demands to create jobs there. WfW has received criticism from the scientific community that job creation has been prioritized too heavily over the objective of improving water resources (van Wilgen and Wannenburg 2016). In addition, the areas that are cleared are often not cleared effectively, due to the technical difficulty and physical demands of invasive species control, for which WfW teams are often insufficiently trained and lack adequate performance incentives.

Key Lessons Learned

WfW's dual goals of invasive species control and poverty alleviation are both the key to its political success and a challenge in its implementation. The level of funding and effort that WfW has garnered for invasive species control could not have been mobilized without its job-creation component. However, when it comes to deciding which areas to prioritize for WfW projects, there is often a tension between allocating funding to those places that are a priority for job creation and those that are a priority for improving water resources. Because WfW is fundamentally a public works program, the job creation priorities take precedence in practice. In addition to the critical component of poverty alleviation, credible science underpinning the policy has been crucial, demonstrating that there is significant alignment of priority places for job creation and for improving water resources, even though there will always be tension over trade-offs where alignment is not perfect. Finally, having champions in government at an opportune time also enabled the success of WfW.

Conclusions

While the government subsidy mechanism can be used to support the provision of many different environmental benefits, it is no coincidence that the examples in this chapter all include water quality or quantity as a main objective. Government institutions are often well positioned to overcome the challenge of high transaction costs by aggregating funds on behalf of diffuse beneficiaries, for example, through taxes or utility rate fees, and to compensate service providers. This mechanism is thus particularly suitable for hydrological services—as illustrated by the three case studies in this chapter as well as in examples from China and Costa Rica in chapters 12 and 13—and for climate stabilization services (see the examples of REDD+ in chapter 11), sandstorm control services (chapter 12), as well as biodiversity conservation (chapters 12 and 13 especially). Because these benefits are often public goods, it is appropriate that public funds pay for their provision.

For these kinds of services, there are often many different beneficiaries who depend on the actions of many landholders or stewards. In the case of Working for Water in South Africa and the Conservation Reserve Program in the United States, it is the national government that aggregates and distributes funds. However, the same principle can be used at different levels of government, such as between the City of New York and the communities in its source watersheds, in the Catskills example, and even between national governments, again in the case of REDD+ in chapter 11 and other forest carbon and biodiversity investments (chapter 13).

The examples in this chapter also illustrate different ways in which these government subsidy programs can incorporate considerations of social equity and inclusion. South Africa's Working for Water explicitly combines the dual goals of securing water resources and providing job opportunities to the rural poor in South Africa. The two US examples also include dimensions of improving rural livelihoods. Conservation Reserve Program funds reduce incentives to produce crops on the least productive, most environmentally sensitive lands, while providing economic support to farmers when crop prices are low. The Catskill Fund for the Future, managed by the Catskill Watershed Corporation, provides grants and loans to promote environmentally sound development and job growth in the communities that provide New York City's water supply. As the Working for Water program in particular highlights, combining environmental and social outcomes can be powerful politically in terms of marshalling funds for the program, but it is not always simple. The most effective places to direct subsidies from an ecosystem services perspective are not necessarily the places that would be prioritized from a social equity perspective. However, the social equity dimensions can be critical for maintaining public and political support.

Key References

Andréassian, Vazken. 2004. "Waters and forests: From historical controversy to scientific debate." *Journal of Hydrology* 291, no. 1-2:1–27.

Appleton, A. F. 2002. *How New York City Used an Ecosystem Services Strategy Carried out Through an Urban-Rural Partnership to Preserve the Pristine Quality of Its Drinking Water and Save Billions of Dollars*. Forest Trends. doi:10.1017/CBO9781107415324.004.

Chichilnisky, Graciela, and Geoffrey Heal. 1998. "Economic returns from the biosphere." *Nature* 391, no. 6668:629.

Coetzer, Anje, and Johann Louw. 2012. "An evaluation of the contractor development model of Working for Water." *Water SA* 38, no. 5:793–802.

Daily, Gretchen C., and Katherine Ellison. 2002. *The New Economy of Nature: The Quest to Make Conservation Profitable*. Washington, DC: Island Press.

Dunlap, David W. 2015. "As a plant nears completion, Croton water flows again to New York City." *New York Times* May 9, 2015, A15.

Ferraro, Paul J. 2009. "Regional review of payments for watershed services: Sub-Saharan Africa." *Journal of Sustainable Forestry* 28, no. 3-5:525–50.

Gorgens, A. H. M., and B. W. Van Wilgen. 2004. "Invasive alien plants and water resources in South Africa: Current understanding, predictive ability and research challenges: Working for Water." *South African Journal of Science* 100, no. 1-2:27–33.

Hansen, LeRoy. 2007. "Conservation reserve program: Environmental benefits update." *Agricultural and Resource Economics Review* 36, no. 2:267–80.

Hellerstein, Daniel M. 2017. "The US Conservation Reserve Program: The evolution of an enrollment mechanism." *Land Use Policy* 63:601–10.

McConnachie, Matthew M., Richard M. Cowling, Charlie M. Shackleton, and Andrew T. Knight. 2013. "The challenges of alleviating poverty through ecological restoration: Insights from South Africa's 'Working for Water' Program." *Restoration Ecology* 21, no. 5:544–50.

National Research Council. 2000. *Watershed Management for Potable Water Supply: Assessing the New York City Strategy*. Washington, DC: National Academies Press.

Schechinger, Anne Weir, and Craig Cox. 2017. *"Retired" Sensitive Cropland: Here Today, Gone Tomorrow?* Washington, DC: Environmental Working Group.

Soil and Water Conservation Society and Environmental Defense Fund. 2008. *Conservation Reserve Program (CRP) Program Assessment*. Ankeny, IA, and New York, NY. http://www.swcs.org/documents/filelibrary/CRPassessmentreport_3BEFE868DA166.pdf.

Turpie, J. K., Christo Marais, and James Nelson Blignaut. 2008. "The Working for Water programme: Evolution of a payments for ecosystem services mechanism that addresses both poverty and ecosystem service delivery in South Africa." *Ecological Economics* 65, no. 4: 788–98.

USDA. 2015. *Conservation Reserve Program Conservation Fact Sheet*. https://www.fsa.usda.gov/Assets/USDA-FSA-Public/usdafiles/FactSheets/archived-fact-sheets/consv_reserve_program.pdf.

Van Wilgen, B.W., P. R. Little, R. A. Chapman, A. H. M. Görgens, T. Willems, and C. Marais. 1997. "The sustainable development of water resources: History, financial costs, and benefits of alien plant control programmes." *South African Journal of Science* 93, no. 9:404–11.

Van Wilgen, Brian W., Greg G. Forsyth, David C. Le Maitre, Andrew Wannenburgh, Johann DF Kotzé, Elna van den Berg, and Lesley Henderson. 2012. "An assessment of the effectiveness of a large, national-scale invasive alien plant control strategy in South Africa." *Biological Conservation* 148, no. 1:28–38.

Van Wilgen, Brian W., and Andrew Wannenburgh. 2016. "Co-facilitating invasive species control, water conservation and poverty relief: Achievements and challenges in South Africa's Working for Water programme." *Current Opinion in Environmental Sustainability* 19:7–17.

Wannenburgh, Andrew. 2018. "Historical expenditure, clearing and employment data." Accessed November 19, 2018. https://sites.google.com/site/wfwplanning/Home/WfW%20historical%20figures.xls?attredirects=0&d=1.

CHAPTER 7

Regulatory Mechanisms

Lisa Mandle, Rick Thomas, and Craig Holland

To ensure the societal benefits of development are not undercut by environmental harms, many governments require developers to offset the environmental impacts caused by construction. These offsets take the form of protection, restoration, or enhancement of habitats, ecological function, or ecosystem services. In recent decades, government regulators have allowed these offsets to occur away from the development site, enabling larger offset projects that are more effective in terms of both cost and environmental benefits. Regulatory-driven mitigation is now the fastest growing source of payments for ecosystem services. Here we present four case studies of regulatory-driven mitigation programs: stormwater mitigation in Washington, DC; nationwide wetland mitigation banking in the United States; biodiversity conservation banking in California; and forest offsets to mitigate greenhouse gas emissions as part of California's cap-and-trade program. These programs result from regulations at multiple levels of government, including the US Clean Water Act, California's Environmental Quality Act and Endangered Species Act, and the California Global Warming Solutions Act. These programs also create business opportunities by allowing those who secure or enhance environmental benefits to sell credits to those who need to meet mitigation requirements. We identify a number of common elements important to achieving success across mitigation programs. These include the selection of effective metrics to measure mitigation outcomes, consideration of the social equity implications of mitigation, and ensuring sufficient demand for credits, especially when markets are new.

Around the world, development activities provide benefits such as new buildings for homes and jobs, and roads enabling access to markets and services. However, this development can also cause the loss of natural capital and biodiversity, with forests cut down, grasslands paved over, and wetlands filled. To guide development in a way that furnishes a net benefit to society, government regulations in the United States often require that projects—both public- and private-sector—undertake a two-step process. First, development projects should avoid harm by minimizing their environmental impacts. Second, projects must offset unavoidable impacts through mitigation, by protecting, restoring or enhancing habitats, ecological function, or ecosystem services.

Initially, compensatory mitigation occurred project-by-project, often on the same site as the development project. In practice, this approach proved both ineffective and expensive. Over time, regulators introduced innovations that increased the size of mitigation projects and created market opportunities for entrepreneurs

In this chapter, we present four contrasting cases studies to illustrate a range of regulatory-driven mitigation mechanisms: green infrastructure for stormwater mitigation in Washington, DC; wetland mitigation banking across the United States; biodiversity conservation banking in California; and forest offsets as part of California's greenhouse gas cap-and-trade program. These examples all share a common underlying mechanism (fig. 7.1).

In order to receive a permit for development or redevelopment, laws require developers to mitigate harms to biodiversity or ecosystem services with an overall goal of "no net loss"—or even sometimes enhancement—of environmental values. The following case studies highlight the range of environmental values that can be addressed through such a mechanism, and the different participatory roles of government, businesses, and civil society organizations.

Case 1. Green Infrastructure Investment for Stormwater Management in Washington, DC

The Problem

As the city of Washington, DC, has grown, natural areas, green spaces, and empty lots have been replaced with buildings and pavement. Now, 43 percent of the city is covered by paved and impervious surfaces that prevent rainfall from being absorbed into the soil. As a result, large amounts of stormwater flow through municipal storm sewers directly into local rivers, carrying oil, sediment, and other pollutants—whatever washes off the streets. Moreover, the

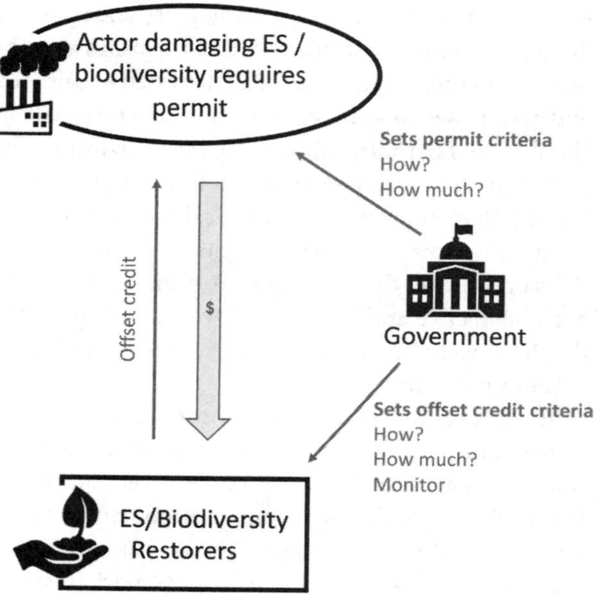

Figure 7.1. In the case of regulatory-driven mitigation, actors who damage ecosystem services or biodiversity pay other entities who restore those losses, in exchange for mitigation or offset credits. The government sets the amount and type of offset credits required of permittees, as well as amount and type of activities that restorers must undertake to generate credits. The government is also involved in monitoring compliance to ensure that credits adequately represent the environmental values intended, whether in terms of ecosystem services, ecological functions, or biodiversity.

city's combined sewage/stormwater runoff system overflows regularly, sending an additional 10 million m³ of runoff and sewage into the Potomac and Anacostia Rivers each year (Holland 2016). This combined flow harms the water quality and ecosystems not only of DC's local rivers, but also of the Chesapeake Bay—the largest estuary in North America—into which these rivers flow (fig. 7.2).

Federal regulators at the US Environmental Protection Agency (EPA) made it clear that DC's stormwater runoff was violating the federal Clean Water Act. To avoid large fines and other penalties, the city was mandated to manage its stormwater. DC's Department of Energy and Environment (DOEE) issued new stormwater regulations in 2013. These regulations required all new developments and major redevelopments in the District to mitigate their impacts on stormwater flows. These new regulations also created the Stormwater Retention Credit market, which allows projects to meet some of the regulatory requirements through the purchase of credits from others who have already implemented nature-based solutions such as green roofs, grass filter strips, wetland construction, and tree planting.

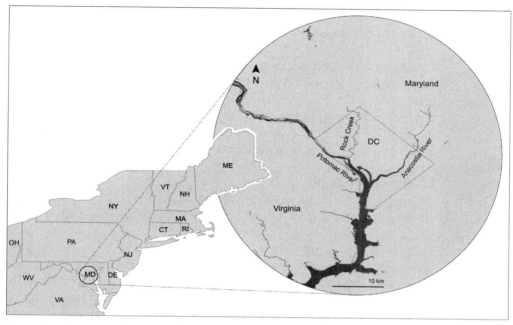

Figure 7.2. Stormwater from the city of Washington, DC—along with debris and pollutants it picks up along the way—runs into Rock Creek and the Anacostia and Potomac Rivers, eventually flowing south into the Chesapeake Bay. Washington, DC, is using green infrastructure, such as green roofs, grass filter strips, and rain gardens, to reduce this runoff and the associated degradation of water quality.

The Ecosystem Service

In the absence of stormwater management infrastructure (either nature-based "green" infrastructure or engineered "gray" infrastructure), stormwater that lands on roofs and paved surfaces picks up sediment, oil, and other pollutants and washes them into sewers, which then flow to DC's three rivers and on to the Chesapeake Bay. When high volumes of water end up in streams after storms, this erodes stream channels and the sediment is carried downstream, degrading aquatic habitat. Green infrastructure therefore contributes two main ecosystem services: improved water quality through filtration of pollutants, and regulation of water flows by increasing infiltration and slowing storm flows. Green infrastructure for stormwater management also provides cobenefits that gray infrastructure does not, including aesthetic values, recreational opportunities, and associated health benefits.

Ecosystem Service Beneficiaries
The improved water quality in DC rivers provides benefits to many DC residents and visitors. Improved water quality in the Chesapeake benefits those involved in recreational and commercial fishing and water-based tourism in the area. The communities located near the mitigation sites are also expected to receive benefits associated with increased green space.

Ecosystem Service Suppliers
The property owners who replace impervious surface with permeable, green cover that contributes to stormwater retention supply the ecosystem services.

Terms of the Exchange: Quid Pro Quo
Under DC's stormwater regulations, major development and redevelopment projects are required to retain a certain amount of stormwater, based on the size of the project footprint. Up to 50 percent of the required stormwater retention can be met through off-site activities; the rest must be retained on-site. DOEE is considering allowing certain developments to petition to use off-site activities for all of their required retention capacity.

A regulated project can meet off-site retention requirements by generating its own Stormwater Retention Credits (SRCs) away from the project site, buying SRCs, paying in-lieu fees, or through some combination of these options. Thirteen preapproved Best Management Practices (BMPs) can be used to meet stormwater retention requirements, or to generate SRCs (DDOE and CWP 2013). Many BMPs, such as green roofs, impervious surface disconnection with grass filter strips, bioretention, vegetated open channels, construction of stormwater ponds and wetlands, as well as tree planting and preservation, include green infrastructure elements.

One SRC is equivalent to retention of ~4 liters of water for one year. Regulated projects that generate their own SRCs implement an appropriate amount of BMPs to meet their off-site retention requirements. Otherwise, projects purchase the SRCs they need from private groups that have implemented BMPs voluntarily. For the in-lieu fee option, fees are paid to the city and go toward developing retention projects constructed and managed by DC's government. The average price of one credit in February 2019 was just over US$2, as compared to an equivalent in-lieu fee of US$3.61. The in-lieu fee is based on the city's estimate of its own costs for implementing BMPs, adjusting annually for inflation. This suggests that the private market is able to provide stormwater retention more cost-effectively than the government, and the city could use in-lieu fees it receives to purchase SRCs at a lower cost than if it implemented retention projects itself. However, the existence of the in-lieu fee option reduces uncertainty about compliance for project developers, should not enough credits be available on the market in the future.

The ability to purchase SRCs provides flexibility to regulated developers, as they can mitigate off-site if it is more cost effective to use the area on-site for additional space or amenities, such as parking spaces. The hope is that this SRC market will also increase both job opportunities and green space in traditionally disadvantaged areas of the city. The cost of land is lower in these areas, so off-site mitigation will likely be cost-effective relative to on-site mitigation in the city's center where most development is happening. Construction and maintenance of off-site mitigation in traditionally disadvantaged areas might also provide employment opportunities.

Between the opening of the SRC market in 2014 and February 2019, over 277,000 credits were traded with a value of over half a million US dollars. The market shows growing activity, with an increasing number of trades and volume of credits sold every year. A similar trading system has been developed in Chattanooga, Tennessee, and has the potential to spread to other cities in the United States (TNC and VCS 2016).

Mechanism for Transfer of Value

Project developers that need SRCs can buy them directly from private SRC-generating projects. Anyone within the city of Washington, DC, can generate SRCs and sell them. Although DC's watersheds extend into neighboring states, trading of SRCs is restricted to its administrative boundaries. Projects must acquire enough credits to cover their stormwater retention needs each year. Credits can be bought on an annual basis or purchased up front and banked for use in future years. SRC sellers are responsible for maintaining SRC-generating projects over time.

A landowner wishing to sell SRCs to a developer must first implement BMPs on their land and receive certification from DOEE. DOEE certifies the number of credits based on the stormwater retention value of the BMPs (see *Monitoring and Verification* for more details). The landowner can then list these credits for sale. Project developers seeking SRCs can contact those selling credits, and the two parties negotiate a mutually agreeable contract, including price, for the sale of credits. They then apply to DOEE to transfer ownership of credits from the seller to the project developer. Upon approval, the project developer takes ownership of the credits and pays the landowner.

As of July 2018, more than 1 million credits were available for sale from twelve groups, ranging from churches to private companies. NatureVest (the conservation investing unit of The Nature Conservancy) and Encourage Capital (a private investment firm), along with an initial US$1.7 million investment from Prudential Financial (a company in the financial services industry), established a company called District Stormwater, LLC. This company is designed to fund, develop, and manage SRC-generating projects, while generating financial returns for investors. District Stormwater's initial projects were successful in generating ecosystem services, cost savings for DOEE, and profit to the company and its investors. District Stormwater is actively pursuing additional projects, as it sees the market as a long-term, profitable business opportunity—and most important, a driver of water quality opportunities in the Chesapeake Bay, and a model for other cities worldwide.

Monitoring and Verification

SRCs are verified and independently audited by DOEE. SRC-generating projects must be designed and installed in accordance with an official Stormwater Management Plan and the DOEE's Stormwater Management Guidebook (DDOE and CWP 2013). Projects must pass a DOEE inspection after being constructed to ensure that BMPs are properly installed, and projects must ensure that BMPs will be maintained. If they meet these conditions, DOEE will award credits for the next three years (*ex ante* crediting), though the site can be reinspected before then to confirm compliance. If the project is not adequately maintained, projects must compensate DOEE for any credits they have already sold—either by buying new credits from elsewhere or paying the in-lieu fee—and lose any unsold credits. Project owners must reapply for credit certification every three years.

DOEE assigns each credit that it issues a unique serial number based on certification year, drainage number, and Stormwater Management Plan. DOEE tracks the ownership of each SRC, along with its use toward mitigation. DOEE must approve transfers of SRC ownership between sellers and buyers to make sure credits are valid. Mitigation projects and credits are listed on an online registry system.

Effectiveness

As of mid-2018, there had not been any formal review or quantification of program performance, which is not surprising given that the credit market only began in 2014. In what was perhaps an indication that the market had gotten a slow start, DOEE announced an SRC Price Lock Program in 2017, which provides US$11.5 million for SRC purchases by the city over a twelve-year contract, or four three-year credit cycles (https://doee.dc.gov/service/src-price-lock-program). The Price Lock Program provides a guaranteed minimum price to credit suppliers for projects located in the most impacted areas of the city's watersheds, though suppliers can choose to sell their credits on the open market if prices are better. The program is a way for DC to incentivize projects that benefit the most critically endangered water bodies, while still allowing the market to remain flexible and simple from a geographic standpoint.

Key Lessons Learned

This case study highlights the critical role of government in enabling market-based mechanisms for financing green infrastructure. The Clean Water Act requirements to limit pollutants entering DC's local waterbodies and the Chesapeake Bay forced the city to implement new stormwater regulation in 2013. Developers who needed permits to manage the stormwater resulting from their projects were forced to choose among engineering stormwater management on-site, paying in-lieu fees, or purchasing SRCs. This, in turn, created the SRC market, generating new business opportunities.

The creation and administration of the SRC Price Lock Program has proven critical for long-term success and durability. When markets are small and lack a history of demand for credits, private investment in credit-generating projects is limited. As a result, publicly administered purchase agreement programs are important to ensure the functioning of new environmental markets until sufficient market volume can generate a clear price signal and demonstrated demand. Government must also create clear guidance on project permitting and tax treatment of credits sales, both of which can slow down market participation. Further, government should support new supply-side entrants to the market in order to get their projects "investment-worthy." DOEE created an

SRC-aggregator start-up grant program to identify potential SRC-generating projects. Predevelopment grants are critical to building a robust supply of businesses able to support the market, and early indications reveal strong demand for these grants. Finally, grandfathering of credits on sites that predate market inception can distort the market and create barriers for new supply-side entrants. While this may provide near-term supply to stimulate demand, it can also create a glut of supply that has no cost basis, sends inaccurate price signals, and discourages investment in new projects.

The challenge now for Washington, DC, is whether a robust market—with an adequate supply of credits meeting the demand from development projects—can be sustained, and whether the program can successfully meet its multiple goals of improved stormwater management with reduced costs to the city and increased benefits to the city's traditionally disadvantaged communities.

Case 2. Wetlands Mitigation Banking in the United States

The Problem

Since the founding of the country, over half the wetlands in the continental United States have been drained, filled, and converted in favor of development, agriculture, and other uses considered to be more economically productive. In the 200 years between 1780 and 1980, wetlands were lost at an average rate of 25 hectares per hour (Dahl 1990). The conversion of wetlands has also meant the loss of wetland-associated ecological functions and ecosystem services that ensure clean water and protect against flooding.

US policy since the late 1980s has called for "no net loss" of wetlands, meaning that any wetland loss due to development should be compensated by comparable wetland restoration, creation, or enhancement. Development projects that impact wetlands are required to mitigate their impacts. Initially, each project mitigated its own impact, often on the project site, or at an off-site location created for that single development project. Mitigation in this form, however, proved to be ineffective. On-site mitigation resulted in small, isolated wetlands whose quality was impacted by the adjacent development, and net loss of wetlands continued. The next stage of improving policy then emerged—wetland mitigation banking—designed to more effectively mitigate losses of wetlands, along with their ecological functions and services. In 2008, under the Clean Water Act, the US Army Corps of Engineers and the Environmental Protection Agency regulations established mitigation banking as the preferred mechanism for mitigating unavoidable impacts to wetlands, as compared to on-site mitigation or in-lieu fees.

The Ecosystem Service

Wetlands provide a diversity of ecosystem services. These include recreational opportunities such as wildlife viewing, boating, and fishing; flood regulation by storing stormwater; water quality regulation by trapping sediment and other pollutants; and protection of coastal areas from inundation and erosion from storms. Mitigation banks create, enhance, or restore wetland ecosystems and their functions as a means of maintaining the ecosystem services they provide.

Ecosystem Service Beneficiaries

As described above, wetlands provide a wide range of ecosystem service benefits shared by much of the public. This includes protection from coastal storms, attenuation of inland flooding, water filtration, groundwater recharge, and recreational opportunities. Wetlands are also important for biodiversity conservation, including supporting migratory bird populations and providing nursery habitat for fish and other aquatic species.

Ecosystem Service Suppliers

The mitigation bank owners and managers who carry out wetland restoration, creation, or enhancement are the ecosystem service suppliers. The mitigation banks are paid by project developers who need credits to mitigate their development of wetlands.

Terms of the Exchange: Quid Pro Quo

Mitigation banks manage land in order to restore, create, or enhance wetland habitat and functions. Officials from the government's Army Corps of Engineers approve the banks, which can sell credits to developers whose activities damage wetlands. Land whose credits have been sold must be protected from development in perpetuity. Corps of Engineers officials also determine the number of credits a developer must buy, based on the size and nature of their impacts on wetlands.

The wetland mitigation banking mechanism benefits developers whose projects impact wetlands. By turning over mitigation activities to mitigation banks run by third parties, developers are able to meet regulatory requirements for compensatory mitigation at a lower cost, and do not have to be responsible for carrying out mitigation themselves when it is not their core business.

Large-scale mitigation banking began in the 1990s, primarily as a means of mitigating impacts by state governments, especially departments of transportation that were building roads and needed to mitigate large wetlands losses. Early banks were generally government sponsored. Mitigation banking expanded rapidly after federal guidance was issued in 1995. By 2001, there were over 200 mitigation banks. Nearly two-thirds were privately owned enterprises. As of July 2018, more than 1,300 mitigation banks have been approved to sell credits for mitigation under the Clean Water Act, with an additional ~300 banks pending (fig. 7.3). Over 200 banks have sold out of credits.

Mechanism for Transfer of Value

Section 404 of the Clean Water Act is the regulatory driver of wetland mitigation banking in the United States. Both private and government development projects that would fill in wetlands or impact streams or other legally eligible aquatic resources are required to first demonstrate that the project minimizes its negative impacts, and second, compensate for (or mitigate) unavoidable negative impacts, in order to receive permits for development. The US Army Corps of Engineers and the Environmental Protection Agency are the two federal agencies that govern permitting authority, though this role has also been devolved to some state agencies. To comply with permits, a project developer may purchase the required number of mitigation credits directly from a mitigation bank or conduct the mitigation itself.

Each mitigation bank, which may be operated by private companies, nongovernmental organizations or the government, is approved by regulators to sell a certain number of credits. Mitigation banks are committed to maintain their wetlands in perpetuity through the use of easements (see chapter 8 for more information on conservation easements). Credits are generally defined in terms of wetland area, rather than using measures of ecosystem service provision. Often, for every hectare of wetland credit a bank is approved for, the bank is required to restore, enhance, or create more than one hectare of wetland. This mitigation ratio (e.g., requiring 5 hectares of wetland mitigation for 4 hectares of credits) is meant to ensure that mitigation adequately compensates for the functions and values of wetlands that are lost due to development, serving as a form of insurance against the risk of failure and differences in wetland quality between the impacted site and mitigated site.

Each bank has a designated service area, which is the geographic area (usually a watershed) in which credits from the bank can be used as mitigation for

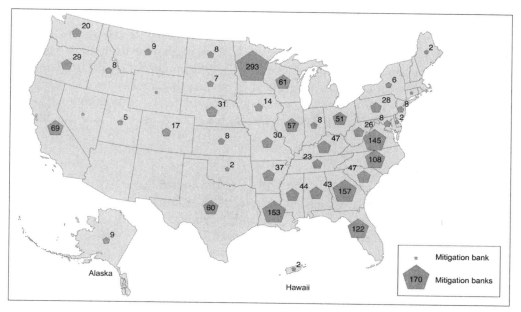

Figure 7.3. More than 1,300 wetland-mitigation banks throughout the United States have been approved to sell credits, and over 200 of these banks have sold out of credits. The label indicates the number of approved mitigation banks per state in states with more than one bank.

impacts. Therefore, credit trading can generally only occur between development projects and mitigation banks within the same watershed. Mitigation banks are listed publicly on the Regulatory In-Lieu Fee and Bank Information Tracking System (RIBITS), run by the US Army Corps of Engineers.

To provide a simple example, consider a project developer whose proposed development would fill in 10 hectares. If the developer wishes to fulfill this obligation by purchasing credits from a mitigation bank, they could use the public registry to locate banks with 10 hectares of available credits whose service area includes the project site. The 10 hectares of credit might correspond to 15 hectares of restored wetlands at the mitigation bank. The mitigation bank sets the price for its credits. When the project developer locates credits they would like to purchase, they pay the mitigation bank directly and the bank deducts those credits from its ledger. Federal and state regulatory agencies track the use of mitigation credits by developers for permitting compliance. Mitigation banks must also report their credit transactions to regulatory authorities.

Monitoring and Verification

Each mitigation bank is overseen by an Interagency Review Team (IRT), responsible for reviewing, approving, and overseeing the bank. Each bank also has a bank instrument, which establishes the number of credits that bank can sell and the ecological assessment techniques that must be used to verify that the bank is providing the necessary ecological functions. Verification happens before credits are released for sale. The instrument also establishes monitoring requirements.

Effectiveness

The rate of loss of wetland acreage in the United States has slowed, but the country is still experiencing a net loss despite mitigation requirements, based on the most recent National Wetlands Inventory. The particular role of mitigation banking in slowing losses is unclear. This is in part because wetland mitigation under Section 404 of Clean Water Act applies only to wetlands over a certain size. The Clean Water Act has also been interpreted in some places not to apply to isolated wetlands. Therefore, a significant number of both small and isolated wetlands are being lost despite these mitigation requirements (NRC 2001).

In addition, trends in wetland acreage or extent do not necessarily reflect trends in ecological function or ecosystem services. In practice, mitigation has often not compensated for the lost societal values of wetlands. Several studies have shown that in certain areas mitigation banking has led to a transfer of wetlands and their associated services away from urban areas, where a large number of people benefit from the ecosystem services provided, and toward rural areas, where fewer people benefit (Ruhl and Salzman 2006). In some cases, this transfer is associated with a loss of wetlands in poor and minority communities, as well (BenDor et al. 2007). Mitigation that tracks wetland area without considering who benefits from the wetland services can lead to redistribution of ecosystem services, creating winners and losers and potentially exacerbating inequality.

Key Lessons Learned

The example of wetlands mitigation banking again highlights how government regulation can set the stage for market-based mechanisms to flourish, creating business opportunities around securing or enhancing natural capital. At the same time, continued wetland losses and redistribution of wetland-associated ecosystem services illustrate the challenges associated with ensuring mitigation adequately compensates for the impacts of development.

Case 3. Conservation Banking in California

The Problem

Inspired by wetlands mitigation banking, the state of California created its own conservation banking program for threatened and endangered species and habitats. Started in 1995, it was the first of its kind in the United States. As with wetlands mitigation banking, conservation banking is designed to pool resources from multiple development projects to fund more effective conservation of larger and more strategic sites that can offset the loss of habitat.

The Ecosystem Service

Conservation banking aims to conserve or enhance threatened or endangered species and habitats. The benefit is biodiversity conservation, which is not an ecosystem service itself but does underpin a range of ecosystem services.

Ecosystem Service Beneficiaries

Biodiversity can be considered a public good, providing broad societal benefits. The particular beneficiaries from conservation banking vary with the specific ecosystem service considered. For example, many people value the continued existence of wild populations of threatened and endangered species, regardless of whether they personally have a chance to see them. Conservation banking can also enhance recreational opportunities, both at conservation bank sites and in adjacent areas where wildlife populations have been bolstered by conservation bank activities.

Ecosystem Service Suppliers

The landowners who create conservation banks are the ecosystem service suppliers. These may be individuals, companies, or nonprofit organizations.

Terms of the Exchange: Quid Pro Quo

The terms of the exchange for conservation banking in California are largely the same as for wetland mitigation banking covered in the previous case study. Conservation banks generate credits by taking actions to protect threatened or endangered species or habitats in perpetuity, such as through conservation easements. Developers whose activities unavoidably harm those species or habitats must offset their impacts through mitigation and can fulfill regulatory requirements for mitigation by buying credits from a conservation bank. The conservation banks are paid when project developers purchase credits. As with wetland mitigation banking, the number of credits that developers are required to purchase and that conservation banks can offer is generally determined by the area of habitat.

One notable difference between conservation banking and wetland mitigation banking, however, is that conservation banking has predominantly relied on conservation of *existing* habitat, rather than restoration, enhancement, or creation of *new* habitat. This means that conservation banking in California serves to stem the loss of biodiversity by increasing the amount of area under protection. It is not designed to, nor will it, achieve no net loss of species or habitats.

Developers or project proponents benefit from being able to meet regulatory compliance at a lower cost and by shifting responsibility for mitigation to conservation banks with expertise in this area. Similar to wetland mitigation banking, the developers pay to offset their environmental impacts by buying credits from conservation banks.

As of July 2018, there were eighty-two state-approved banks overseen by the California Department of Fish and Wildlife (fig. 7.4). Approximately half of these are conservation banks, offering credits for species and habitats. The remaining mitigation banks may offer credits both for wetlands mitigation and for species and/or habitats. Applications to the state for new banks has declined since 2013, with only sixteen applications in 2016. Since its start in California in 1995, conservation banking has spread across the United States. In 2003, the US Fish and Wildlife Service issued federal guidelines for conservation banking. As of 2016, 138 banks had been federally approved, covering nearly 80,000 hectares, with more than 50 percent of banks located within California (Gamarra and Toombs 2017).

Mechanism for Transfer of Value

Conservation banking in California functions much in the same way as wetlands mitigation banking. Because conservation banking is a state- rather than federal-level program, the government agencies involved differ. The California Environmental Quality Act and the California Endangered Species Act provide the regulatory drivers of conservation banking. They require compensatory

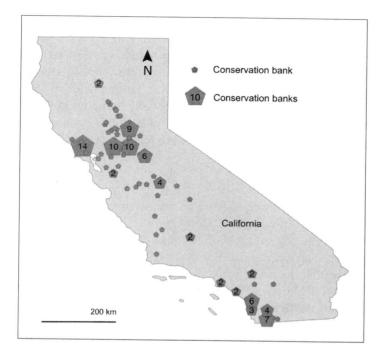

Figure 7.4. Location of conservation banks in California with credits for mitigating impacts to species and/or habitats. These include the 82 banks approved by the California Department of Fish and Wildlife mentioned in the text, as well as banks approved by other agencies. In areas where multiple banks are closely clustered, the number of mitigation banks in that area is labeled.

mitigation for projects that "substantially diminish habitat for fish, wildlife or plants" or impact threatened or endangered species. The California Department of Fish and Wildlife reviews and approves conservation banks, determines the number of credits designated, and oversees and monitors bank operations. A list of banks and available credits are maintained online.

A developer who needs to mitigate impacts to species or habitats in order to receive regulatory approval for their project can seek available conservation bank credits that match their needs. The developer then buys the credits from the conservation bank and the bank deducts those credits from its ledger. With approval of the transaction by regulatory authorities, the developer can then receive the permits needed to proceed with their project.

Monitoring and Verification
The California Department of Fish and Wildlife verifies the number of credits a bank can offer before credits are released for sale. As mentioned previously, this is generally based on the area of habitat, or area of occupied habitat for

a particular species. It may also be based on the number of breeding pairs. The same agency also monitors compliance after credits have been issued. Monitoring tends to include only annual or seasonal surveys of species or habitat-specific factors (e.g., water level in vernal pools) for those species or habitats for which the bank was established. However, especially for species with large ranges, this kind of monitoring is not especially useful for evaluating the success of the bank because of the challenge of attributing changes in species abundances to the bank itself as compared to regional factors beyond the bank's boundaries.

Effectiveness

Conservation banks are generally considered an improvement over project-by-project and on-site mitigation. A 2005 study reported that 49 percent of conservation banks, protecting over 11,000 hectares of habitat, would have been destroyed or seriously degraded if not otherwise protected by conservation banking, suggesting that some banks do provide additional conservation value (Fox and Nino-Murcia 2005). A more recent study confirmed that species conservation banks reduce habitat loss relative to similar areas outside of banks (Sonter et al. 2019). However, the same study also found higher restoration rates outside of banks, complicating an understanding of banks' net contribution to California's conservation goals. Indeed, many banks are still effectively stand-alone efforts. Regional conservation plans ideally would guide the location of banks to promote connectivity and other conservation objectives. However, development of such plans has been slow.

Bank proposals increased year over year from 1995 to 2012, but they have declined annually since then. No new banks were approved between 2009 and 2013. There are a number of likely reasons for this. A lengthy approval process (two to seven years), due in part to staff and funding shortages, proved a deterrent; new banking standards effective since 2013 have shortened timelines (Bunn et al. 2013). The California Department of Fish and Wildlife began a new fee system in 2013, with the aim for banks and project developers purchasing credits to fully fund the associated government administrative costs. As of 2017, these fees totaled around US$100,000 across multiple stages of review and implementation. Conservation bankers also report challenges in assessing the costs and financial risks of proposed banks up front. As of 2005, 35 percent of for-profit conservation banks reported that they had broken even. Although these numbers are dated, they suggest that profitable operation of conservation banks can be a challenge.

Key Lessons Learned

California's conservation banking program illustrates how mitigation banking can be extended beyond wetlands to promote preservation of species and habitats. It also highlights the challenges in creating a self-sustaining program that provides sufficient certainty and financial incentives for private entities to create banks, coupled with sufficient resources for the government to provide the necessary review and oversight.

Case 4. Carbon Markets in California

The Problem

Climate change poses a number of threats to the health and productivity of California's residents. The state is expected to experience hotter and more prolonged heat waves, with inland areas already prone to drought experiencing the most severe effects. These heat waves, alongside predicted declines in air quality (meaning more days that exceed the federal standards for ozone) will result in adverse effects not just on the human population but on agricultural production, as well—a US$40 billion industry in California. The state is also at greater risk of wildfire and sea-level rise, both of which pose numerous threats.

Within the United States, California is second only to Texas in state-level greenhouse gas emissions. The top three sources of emissions in California are transportation, electric power, and industrial use, together accounting for over 85 percent of the state's total emissions (ARB 2014). Given that these sectors are critical for California's growth and development, the problem the state faces is how to reduce its emissions—and therefore its future risk—without compromising its economy.

To date, the state has adopted an aggressive climate policy. The state's governors have issued a series of executive orders aimed at reducing greenhouse gas emissions. As of 2018, the most ambitious is Executive Order (EO) B-55-18, which sets a goal of statewide carbon neutrality by 2045. In addition, Senate Bill (SB) 100 commits the state to clean electricity by 2045.

California's most expansive climate policy came into effect in 2006 with the passage of Assembly Bill (AB) 32 or the "California Global Warming Solutions Act." This bill limits the state to 1990 emission levels by the year 2020. Part of the state's approach to achieving this is through a cap-and-trade market for greenhouse gas emissions, targeting the state's biggest polluters. Under this program, the state's Air and Resource Board (ARB) sets a limit on total

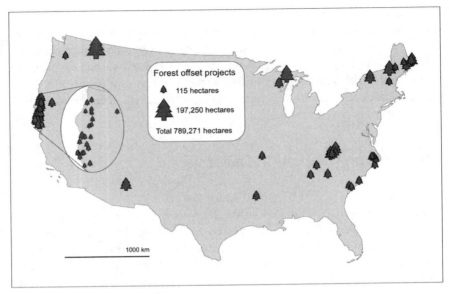

Figure 7.5. Forest offsets projects under CA AB-32. Close to 800,000 hectares within 59 forest management projects across the continental United States have been approved to sell offset credits under California's cap-and-trade market for greenhouse gas emissions.

emissions, allowing companies to decide how to comply with the regulation, whether by decreasing their emissions, purchasing allowances, or investing in offset credits.

Offset credits can be purchased for verified emission reductions from a number of different initiatives, including forest management, elimination of ozone depleting substances, livestock management, methane capture from mines, and improved rice cultivation strategies. Here, we focus on offsets purchased through forest management, the source of the majority of available offset credits (figure 7.5).

The Ecosystem Service
The primary ecosystem service provided by offset credits is climate regulation by carbon sequestration. Forest management offset credits specifically can generate additional ecosystem services including water filtration and soil erosion prevention.

Ecosystem Service Beneficiaries

There are two broad groups that benefit from these offsets and the ecosystem services they provide. Because carbon storage affects the global climate system, the world's population is a beneficiary of the climate regulation provided by forest management. Communities located near the forest management areas also stand to benefit, as many of the regulating services forests provide are local.

Ecosystem Service Suppliers

As defined under AB-32's offset program, the ecosystem service suppliers are parties that choose to undertake emission reduction projects and sell the credits they generate to companies regulated under AB-32. These suppliers are referred to as Offset Project Operators (OPOs) and are responsible for ensuring that projects meaningfully reduce emissions. In the case of forest management offsets, OPOs can be businesses, individuals, or partnerships, so long as they are a forest owner and have the legal authority to implement an offset program there. For an OPO to be paid for their project, they must list it on an ARB-approved Offset Project Registry, such as the Verified Carbon Standard, and adhere to all requirements listed in ARB's Compliance Offset Protocol (discussed in more detail below). Assuming the project is approved, an OPO is issued ARB offset credits, which are then made available for regulated companies to purchase.

Terms of the Exchange: Quid Pro Quo

The cap-and-trade program covers approximately 350 businesses that together account for 85 percent of California's emissions, primarily electricity generating utilities, electricity importers, large industrial facilities, and fuel distributors. The total limit of statewide emissions (the "cap") declines over time to meet state total emissions targets; the 2013 cap was set at 2 percent below 2012 emissions forecast, declining ~2 percent in 2014 and ~3 percent annually from 2015 to 2020.

At the program's inception in 2012, companies received ~90 percent of their emissions allowances free (one allowance = 1 $MMTCO_2e$). Emissions beyond these free allowances can be covered by purchasing more allowances through auctions or by purchasing offset credits. Because these two options equally meet a regulated company's requirement, the company is expected to purchase whichever is cheaper. AB-32 mandates that the amount of free allowances distributed go down over time, forcing companies to either find innovative ways to reduce their emissions, purchase more allowances, or rely more on offsets. To ensure companies are not simply paying their way into compliance without ever addressing their own emissions, offsets are capped at 4 percent of an entity's emissions levels.

Mechanism for Transfer of Value

Parties interested in selling offset credits must adhere to specific Compliance Offset Protocols ARB has developed covering each eligible sector: forestry, urban forestry, manure digesters, destruction of ozone-depleting substances, capturing and destroying methane from mines, and reducing emissions from rice cultivation. Each protocol has sector-specific requirements for which actions generate credits. For forest management these activities include avoiding conversion of forestland, reforestation, and improving forest management.

All offset protocols stipulate that projects are located within the United States and go beyond "business as usual," meaning offset credits cannot be generated for greenhouse gases these projects were already going to sequester without the purchase of offsets. Half of the offsets must have environmental benefits in California. Once a company has purchased offset credits and used them to meet their emissions reduction requirements, the credits must be "retired" to prevent them from being counted multiple times. In the case of forestry offset credits, there are plans to allow companies to purchase international forest credits in the future. California is currently working with the states of Chiapas, Mexico, and Acre, Brazil, to set up REDD+ initiatives that are compatible with AB-32 requirements to strengthen the offset credit market (see chapter 11 for more information about REDD+).

Monitoring and Verification

Regulated companies must report their total greenhouse gas emissions to the ARB and undergo a third-party verification process to ensure compliance. Offset projects are also subject to a monitoring and verification procedure. For forest management, the protocol mandates that every project undergo site visits, have an extensive tree monitoring plan to quantify forest stocks, and submit reports for independent verification by an ARB-accredited body every six years. These monitoring and verification procedures must continue for 100 years following the distribution of offset credits.

Effectiveness

According to the state's Greenhouse Gas Inventory report, in 2016 California's emissions dropped 13 percent from their peak in 2004. This means that the state has already met its 2020 target of lowering emissions to their 1990 emissions level and is now working toward a 40 percent reduction of 1990 emissions by 2030. Although promising, it is unclear how much to attribute this success to AB-32 alone since it has been accompanied by a suite of other emissions-reductions policies.

A 2017 analysis of all the forest management programs registered with ARB identified thirty-nine projects around the nation, each which have been credited with offsetting 654,000 tons of carbon-dioxide equivalent on average (Anderson et al. 2017). Importantly, the study concluded that the reductions were additional, meaning they contributed beyond what would have occurred without the offset purchases. Although the cap-and-trade program was initially scheduled to expire in 2020, the state legislature approved a ten-year extension in 2017, indicating statewide confidence in the program.

Key Lessons Learned

Implementation of California's cap-and-trade program has produced a number of important insights that can inform similar programs elsewhere. Particularly, the state has succeeded in finding cost-effective ways to administer the program. One of these strategies is to develop partnerships to aid in the costly administrative tasks associated with the program. In 2014 the state linked its cap-and-trade program with Quebec, Canada's own emissions reductions program, simultaneously cutting down on its administrative costs while adding credibility to California's program. Through this partnership the two parties have been able to pool their resources to harmonize regulations and guidance documents, and even operate a joint auction platform to sell emissions allowances.

The state also helps cover program costs by collecting fees from the largest sources of greenhouse gases (approximately 250 fee payers, responsible for 330 $MMTCO_2e$/year). Because these fees are based on the payer's emissions, this policy serves to further incentivize emissions reductions from these sources.

California's success also highlights the importance of a well-designed offset credit market. ARB's market is successful because it balances rigorous requirements (ensuring meaningful reductions) with a relatively streamlined process (lowering transaction costs for potential Offset Project Operators). To date, over 40 million offset credits have been verified from approximately 280 projects. Notably, nearly 80 percent of these credits come from forest management projects. This indicates a willingness among forest owners to participate in offset markets and an opportunity to produce additional benefits beyond carbon storage. In fact, when surveyed, 92 percent of forest owners operating an offset project indicated at least one cobenefit they perceived as a result of their program, including water quality, recreation, and wildlife.

Conclusions

Looking across the case studies presented, several lessons emerge. First, mitigation often aims to maintain a certain level of environmental quality relative to an established baseline, whether measured in terms of biodiversity, ecological function, or ecosystem services. A common challenge is how to measure progress toward this goal, as well as how to design and enforce regulations in order to achieve it. The case studies suggest that progress is easier to track when well-defined metrics appropriate to the end goal are matched with appropriate geographic constraints for mitigation activities. For example, the stormwater regulation services provided by green infrastructure under the Stormwater Retention Credit system in Washington, DC, are measured in terms of the volume of stormwater retained. This is in contrast to wetland mitigation banking, which has relied less effectively on wetland area as a rough proxy for a wide array of wetland functions and services that could be measured directly.

Second, and related, the examples in this chapter highlight the varying impacts mitigation can have on social equity and environmental justice. In some cases, mitigation could produce win-wins between environmental outcomes and equity. For example, Washington, DC's program hopes to increase green amenities in poor, urban neighborhoods in addition to increasing stormwater regulation services. However, wetland mitigation in the United States has had the opposite effect, with the tendency for wetland-based ecosystem services to move away from developed areas where they were benefiting poor and minority communities to more rural areas. All this points to the need to explicitly consider the equity implications of mitigation when designing regulations, to avoid unintended consequences.

Finally, extending mitigation programs to the largest geographic extent possible, without compromising environmental goals or social equity, is helpful for creating a robust market. Linking California's emissions credit market with Quebec's means that those interested in selling credits can enter the market with greater confidence that there will be sufficient demand for offsets from regulated entities. In the case of wetland mitigation banking, privately operated banks have been established by multiple, professional actors with access to private capital. Even though credits can generally be traded only within watersheds, the similarity in systems across jurisdictions as a result of the underlying national policy makes it easier for private firms to scale their approaches. In contrast, because the market for stormwater retention credits in Washington, DC, is only citywide and therefore relatively small, a government-operated purchase agreement program was needed to ensure sufficient demand for credits in the market's early days. In sum, issues of metrics, social equity, and market scale all deserve careful consideration when designing mitigation programs.

Key References

Anderson, Christa M., Christopher B. Field, and Katharine J. Mach. 2017. "Forest offsets partner climate-change mitigation with conservation." *Frontiers in Ecology and the Environment* 15, no. 7:359–65.

ARB (California Air Resources Board). 2014a. *Assembly Bill 32 Overview.* https://www.arb.ca.gov/cc/ab32/ab32.htm.

———. 2014b. *First Update to the Climate Change Scoping Plan.* Sacramento, California: California Environmental Protection Agency Air Resources Board.

———. 2014c. *U.S. Forest Projects Compliance Offset Protocol.* https://www.arb.ca.gov/regact/2014/capandtrade14/ctusforestprojectsprotocol.pdf.

Bayon, Ricardo, Nathaniel Carroll, and Jessica Fox. 2012. *Conservation and Biodiversity Banking: A Guide to Setting up and Running Biodiversity Credit Trading Systems.* London: Earthscan.

BenDor, Todd, Nicholas Brozovic, and Varkki George Pallathucheril. 2007. "Assessing the socioeconomic impacts of wetland mitigation in the Chicago region." American Planning Association. *Journal of the American Planning Association* 73, no. 3:263.

Bunn, David, Mark Lubell, and C. Johnson. 2013. "Reforms could boost conservation banking by landowners." *California Agriculture* 67, no. 2:86–95.

Dahl, Thomas E. 1990. *Wetlands Losses in the United States, 1780's to 1980's. Report to the Congress.* No. PB-91-169284/XAB. National Wetlands Inventory, St. Petersburg, FL (USA).

DDOE and CWP (District Department of the Environment and Center for Watershed Protection). 2013. *Stormwater Management Guidebook.* https://doee.dc.gov/swguidebook.

Fox, Jessica, and Anamaria Nino-Murcia. 2005. "Status of species conservation banking in the United States." *Conservation Biology* 19, no. 4:996–1007.

Gamarra, Maria Jose Carreras, and Theodore P. Toombs. 2017. "Thirty years of species conservation banking in the US: Comparing policy to practice." *Biological Conservation* 214:6–12.

Holland, Craig. 2016. "Financing solutions for storm water run-off." *Environmental Finance.*

www.environmental-finance.com/content/analysis/financing-solutions-for-storm-water-run-off.html.

NRC (National Research Council). 2001. *Compensating for Wetland Losses under the Clean Water Act*. Washington, DC: National Academies Press.

Ruhl, J. B., and James E. Salzman. 2006. "The effects of wetland mitigation banking on people." *National Wetlands Newsletter* 28, no. 2:1, 8–13.

Sonter, Laura J., Megan Barnes, Jeffrey W. Matthews, and Martine Maron. 2019. "Quantifying habitat losses and gains made by U.S. Species Conservation Banks to improve compensation policies and avoid perverse outcomes." *Conservation Letters* e12629 https://doi.org/10.1111/conl.12629.

TNC and VCS (The Nature Conservancy and Verified Carbon Standard). 2016. *Developing a Modular Stormwater Crediting Program to Reduce Stormwater Runoff by Scaling up Construction of Green Infrastructure, Phase 1: Analyzing Existing Initiatives to Inform the Development of a Stormwater Crediting Program*. Report to the David and Lucille Packard Foundation.

CHAPTER 8

Voluntary Mechanisms

Lisa Mandle and Meg Symington

Many ecosystem services on both private and public lands could be increased with greater support. Regulation and direct government funding have limits, thus voluntary support from philanthropically motivated private actors, nongovernmental organizations, and bi- and multilateral organizations contribute to conserving the lands around the globe that provide these services. In this chapter, we describe two examples of voluntary conservation: the Transition Fund for the Amazon Region Protected Areas Program (ARPA), and conservation easements by land trusts in the United States. In the case of ARPA, the Brazilian government is strengthening and expanding its protected area network with funding from a coalition of private foundations, companies, and other governments. In the case of conservation easements, protection of private land is subsidized by land trusts and by tax breaks from the government. Together, these approaches have contributed to the protection of large areas—60 million hectares in the Brazilian Amazon and 7 million hectares within the United States—with diverse ecosystem service benefits. Avoided deforestation within the Amazon protected area network is contributing to reduced carbon emissions and global climate benefits from ARPA sites. Both models have spread to other countries, adapting to varying legal frameworks, as well as cultural and environmental contexts.

In addition to the critical role that governments play in conservation through subsidies and regulatory requirements (see chapters 6 and 7), philanthropically minded individuals, nongovernmental organizations, and private companies can contribute powerfully to conservation through their voluntary actions (fig. 8.1). This chapter presents two examples of voluntary conservation: (1) the Amazon Region Protected Areas program (ARPA) in Brazil, and (2) land

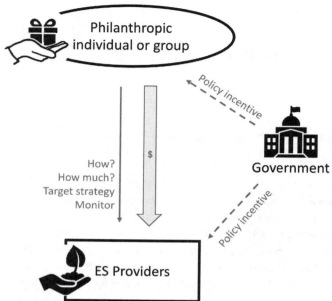

Figure 8.1. With voluntary mechanisms, philanthropically minded individuals or groups provide funding directly to ecosystem service providers. Government may play an indirect role in the transaction by providing incentives to ecosystem service providers and/or to the philanthropic actors.

trusts in the United States. In the case of ARPA, the Brazilian government is the ecosystem service provider, with international funding from private foundations, companies, and other governments providing a catalyst for expanding and strengthening its protected area network. In the case of land trusts, the ecosystem service providers are private landowners, whose actions are subsidized by the government through tax breaks.

Case 1. ARPA (Amazon Region Protected Areas Program) with the Transition Fund

The Problem

The Brazilian Amazon, which includes around 60 percent of the total Amazon forest, has lost nearly 20 percent of its forest cover since 1970, a result of logging, agricultural expansion, and other development. This amounts to a loss of over 750,000 km²—an area larger than the island of Borneo—and poses a threat to globally important biodiversity and climate regulation functions located there, as well as a threat to water resources and life-support systems in the region, and the livelihoods of communities dependent on the forest. In response, the government of Brazil designed protected areas covering nearly 30 percent of the Brazilian Amazon. However, many of these areas are effectively "paper parks," existing in name only and without the funding needed to manage the areas to avert deforestation. This challenge led to the development of the Amazon Region Protected Areas program (ARPA) and Transition Fund, based on the Project Finance for Permanence (PFP) model. This program brings together multi- and bilateral donors, private philanthropic organizations, and the Brazilian government to enact large-scale conservation measures, with the financial resources needed for long-term conservation.

The Ecosystem Service
Securing and expanding the protected area network in the Brazilian Amazon will provide a range of ecosystem service benefits—from local to global climate regulation; to regulation of water flow and quality for hydropower, irrigation, and drinking; to provisioning services to local communities, including food, fiber, and medicinal products. It will also contribute to global priorities for biodiversity conservation. (See the REDD+ cases in chapters 11 and 13 for more on forest-related ecosystem services.)

Ecosystem Service Beneficiaries
Conservation of the Brazilian Amazon provides benefits at multiple scales, from the global beneficiaries of climate regulation (through carbon storage and sequestration) and biodiversity conservation; to the national communities who benefit through improved energy, water, and food security; to local communities whose livelihoods are dependent on the forest.

Ecosystem Service Suppliers
The Brazilian government and communities who manage lands sustainably are the ecosystem service providers in this example.

Terms of the Exchange: Quid Pro Quo
Under the terms of the deal, Brazil will protect 15 percent of the Brazilian Amazon—60 million hectares, an area 1.5 times the size of California, or 3 times the size of all US national parks. Half of ARPA's protected areas are for

strict conservation use, while half are for sustainable use, allowing some forms of extraction.

ARPA donors created a Transition Fund, a long-term sinking fund, designed to cover recurrent costs of ARPA for twenty-five years. The fund has a target of US$215 million, with nearly US$211 million committed as of June 2017. Funding is released from the Transition Fund to pay the government of Brazil for protected area management—including on-the-ground protection (e.g., salaries for fire brigades, and lodging, gasoline, and boats for patrolling), participatory management, equipment, infrastructure and operations, monitoring and research—assuming the government meets certain targets along the way. There are eleven disbursement conditions, which include creating new protected areas, implementing biodiversity monitoring, meeting protected area staffing targets, and securing funding from the government of Brazil.

This Project Finance for Permanence (PFP) model has been used in Costa Rica (see chapter 13) and in the Great Bear Rainforest in British Columbia. Similar programs are also being developed in Peru, Colombia, and Bhutan.

Mechanism for Transfer of Value

Under the PFP model, funding is assembled from multiple donors, but no individual donor is obligated to disburse funds until the fund target is met. Donors make a commitment at the fund's "closing." The ARPA Transition Fund is funded by a mix of philanthropic organizations (Gordon and Betty Moore Foundation, Linden Trust for Conservation, Bobolink Foundation), government and multilateral organizations (World Bank, Global Environment Facility (GEF), the Inter-American Development Bank (IDB), KfW (German Development Bank)), conservation organizations (WWF), and private companies (Anglo-American mining company).

The Transition Fund will be drawn down over time, with its funding disbursed to the ARPA areas as the government meets its agreed-upon conservation outcomes and financial targets. At the same time, the government of Brazil must increase the amount of funding it provides, until it completely takes over funding of its protected areas by 2038.

The ARPA vision was launched in 2002 at Rio+10, with closing for its financial package occurring in May 2014. From 2002 to2009, the program focused on creating new protected areas. From 2010 to 2017, new and existing protected areas were "consolidated" (they met certain benchmarks for management and monitoring) within the ARPA system. Finally, starting in 2014,

ARPA is shifting from being funded by the Transition Fund to being 100 percent funded by the Brazilian government. By the end of transition funding in 2038, the Brazilian government expects to support protected area management largely through environmental compensation funds paid by companies to offset impacts of infrastructure development (see chapter 7 on regulatory-driven mitigation, and especially in-lieu fees), as well as payments for ecosystem services.

Monitoring and Verification

The Transition Fund Committee, which includes representatives from donor institutions and the Brazilian government, reviews the status of disbursement conditions and approves biannual disbursement from the fund, assuming conditions are met. As mentioned previously, development and implementation of a biodiversity monitoring program is one requirement for disbursement of funds. A standardized questionnaire and datasheets—the Protected Area Management Effectiveness Tool—is also used to track whether protected areas meet achievements for twelve indicators, including land tenure, existence of a management plan, participatory management, and monitoring.

Effectiveness

ARPA has met most of its targets, although the federal government has faced challenges designating new protected areas. In 2017, three existing protected areas covering nearly 1.5 million hectares were brought into the ARPA system, bringing the total size of ARPA to 117 protected areas and 60.8 million hectares (figure 8.2). While ARPA now includes the targeted 60 million hectares of protected areas, the creation of new protected areas within that total is lagging. ARPA was targeted to create 6 million hectares of new protected areas as one of the disbursement conditions for the Transition Fund. As of mid-2018, an additional 3 million hectares of new protected areas were still needed to meet this goal. The Transition Fund Committee has extended the deadline for the addition of new protected areas until the end of 2019. The government also fell short of meeting its protected area staffing targets for 2017. Based on this, disbursements for 2018–2019 were penalized 3.5 percent.

The ecosystem service benefits of ARPA could be large. Based on the observed effect of ARPA on deforestation between 2003 and 2008, a 2010 study estimated that the full expansion of ARPA would prevent the release of 1.4 Petagrams carbon emissions, equivalent to ~16 percent of global greenhouse

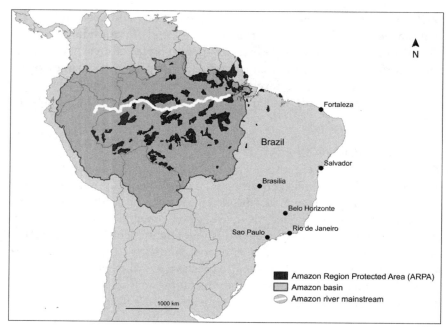

Figure 8.2. The Amazon Region Protected Areas (ARPA) system includes more than 100 protected areas in Brazil, covering over 60 million hectares. Note that the ARPA areas that are outside the Amazon Basin are within the Brazilian Legal Amazon. Data courtesy of WWF Brazil.

gas emissions per year (Soares-Filho et al. 2010). An updated study calculated that ARPA-supported protected areas prevented ~0.35 Petagrams. of carbon emissions between 2005 and 2015 (Soares-Filho 2016).

On the financial side, ARPA experienced some challenges in securing the delivery of funds committed at closing in 2014, though it has commitments for US$211 million in funding out of its US$215 million target. Funding from GEF and Anglo-American (a mining company) was later to come in than originally projected. The large expected contribution from Brazil's Amazon Fund is also of concern, especially given Norway's reduction in its 2017 allocation to the Amazon Fund due to rising deforestation rates in the Brazilian Amazon (see chapter 11 for more about the Amazon Fund). On the other hand, the decline in the value of the Brazilian real (from 2.2 BRL to the US$ in 2014 to 3.8 in August 2018) has resulted in more funding than was anticipated when the Transition Fund financial model was developed. The Brazilian government has also had challenges spending the funding that has been released so far toward consolidation of protected areas, with a lower rate of spending than originally planned. State governments have faced challenges in complying with the financial reporting requirements of ARPA and, in the future, could face difficulties securing counterpart funding needed to continue to receive funds from the Transition Trust. Finally, recent political changes—including weakening of environmental legislation and attempts to eliminate or downgrade existing protected areas—and economic instability in Brazil have created new challenges.

Key Lessons Learned

The example of ARPA and the Transition Trust illustrates the power of the Project Finance for Permanence model. By bringing together contributions from multiple donors from the private and public sectors at once, it is possible to fund significant, coordinated conservation action at a national scale. This would not be possible with funds that arrive bit by bit over time. The model is most likely to succeed at assembling a broad coalition of donors when applied to areas with high value to a wide array of actors, such as areas with threatened carbon stocks and high biodiversity of importance internationally.

Two other aspects of the ARPA example that have contributed to its success so far are worth highlighting. First is the conditioning of the disbursement of funds on meeting predefined targets. This pay-for-performance approach makes participation less risky to donors. Second is the design of the Transition Trust as a sinking fund that will be drawn down over time as Brazil takes on increasing responsibility for funding. This provides an exit strategy for donors, while giving the Brazilian government time to ramp up its own capacity and funding. With these features, it is easy to see why the PFP model is an attractive option for financing conservation from the perspective of both donors and governments.

Case 2. Land Trusts and Conservation Easements in the United States

The Problem

Over 70 percent of land in the United States is privately owned, nearly ten times the amount of land in IUCN-categorized protected areas within the continental United States. Much of this privately held land has substantial environmental value. What incentives can be provided to private landowners to secure these public benefits? The land trust and conservation easement mechanism emerged in the 1980s in the United States as a way for private landowners to voluntarily and permanently restrict development on their land to secure its environmental values in exchange for tax incentives from the federal government. According to a 2015 national land trust census (Land Trust Alliance 2016), there are over 1,300 active land trusts in the United States protecting nearly 7 million hectares with conservation easements (fig. 8.3).

The Ecosystem Service

Conservation easements can be used to secure a wide variety of environmental benefits. This includes protecting water resources or other ecosystem services, as well as protecting biodiversity, providing corridors between or buffers to protected areas, and also conserving working landscapes such as farms and ranches.

Ecosystem Service Beneficiaries

Conservation easements are required to provide some public benefit in terms of natural, cultural, or historic value, in order for the landowner to receive tax benefits. The beneficiaries of a particular easement depend on the landscape context of the easement, management of the land, and whether direct access to the land is needed to enjoy its benefits.

Ecosystem Service Suppliers

The private landowners are the suppliers of ecosystem services or other environmental benefits, and they are compensated through tax benefits. Land trusts contribute to managing easement lands for those benefits and operate as nonprofit corporations also eligible for tax benefits.

Terms of the Exchange: Quid Pro Quo

Land trusts are private, nonprofit organizations that acquire and steward land for conservation. When a private landowner wants to place a conservation easement on their land, they enter into an agreement with a land trust. The conservation easement is attached to the property title, permanently restricting future land uses. For example, the conservation easement may prevent future development or subdivision of the land. The private landowner then transfers the easement to the land trust, either by selling the easement or donating it. If

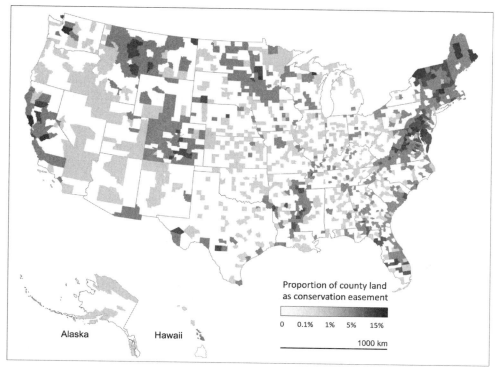

Figure 8.3. Conservation easements are widespread across the United States.

the easement is donated and the land has less value as a result, the donation counts as a charitable contribution and the owner can receive income tax deductions up to the loss in value. Even if the easement is sold, the restrictions on future development may reduce the assessed value of the land, which can reduce property taxes for the owner as well.

While the landowner transfers the easement to the land trust, the land itself is still the property of the original owner, who can later sell the land or pass it on to their heirs. The easement is permanent, however, and travels with the title. As a result, subsequent owners must continue to comply with its terms.

Mechanism for Transfer of Value

A conservation easement is a voluntary, legal contract between a landowner and a land trust or government agency. The easement permanently restricts uses of the land to protect its natural capital values. Land trusts may acquire land directly ("in fee simple") or acquire conservation easements. Land trusts vary greatly in terms of the amount of land they hold and their goals in acquiring land. Some land trusts operate at the state or national level, while others focus much more locally. Land trusts may target land based on its value for biodiversity or key species, or may focus on preserving the cultural values associated with working landscapes.

For the conservation easements they hold, land trusts ensure that the land is managed in compliance with conservation objectives and the land-use restrictions specified in the easement terms. The particular restrictions associated with a given conservation easement depend on the conservation goals the easement is designed to secure, the mission of the land trust, and the needs of the landowner. They typically include restrictions on subdividing the property; developing it for housing, commercial, or industrial use; or limitations on activities such as mining or housing development that would damage habitat or reduce ecosystem service provision.

The conservation easement mechanism primarily involves an exchange between a private landowner and a private non-profit organization (a land trust). Government plays a substantial though indirect role, however, by providing a number of tax benefits that subsidize these transactions. The federal government has also provided funding to land trusts for easement acquisition through a variety of programs, including the US Forest Service's Forest Legacy Program and the US Fish and Wildlife Service's National Fish and Wildlife Foundation. Conservation easements may also be transferred to the government, rather than to a land trust.

Monitoring and Verification

The holder of the easement (usually land trusts) is responsible for monitoring easement compliance. A survey of conservation easements held by The Nature Conservancy (TNC), the largest land trust in the United States, found that TNC routinely audited its own easements for legal compliance with restrictions on development (92 percent within the past three years), but it quantitatively monitored fewer than 1 in 5 conservation targets. There is generally no public accountability or third-party verification of compliance. Effective monitoring, especially of conservation objectives, therefore remains a concern.

Effectiveness

The land trust mechanism has been successful at acquiring conservation easements across a substantial area of land in the United States. According to a 2015 land trust census (Land Trust Alliance 2016), there are 1,362 active land trusts, conserving over 22 million hectares through a variety of mechanisms, including nearly 7 million hectares under easement, as well as direct acquisition of land, and assisting with transfer of land to government agencies or other organizations. This represents nearly a threefold increase from ~2.5 million hectares under easement a decade earlier in 2005. In 1985, the earliest year with available data, less than 400,000 hectares were under easement.

While the area under conservation easements is impressive, there has not been a systematic accounting for the values provided by these lands. The ecosystem service or biodiversity benefits that result from establishing conservation easements therefore are difficult to quantify. A study in the state of Wyoming found that in areas with high development pressure, there was reduced development on land with easements as compared to without over a five-year period, as well as higher utilization by some wildlife. However, wildlife use of a given site was affected more by the environment in the surrounding area beyond the site rather than by a site's easement status, reflecting the limitation of the easement approach for mobile wildlife. The conservation easement approach has also been criticized as being too haphazard for effective conservation. TNC's review of its easements concluded that easement acquisition was generally strategic and increasingly so over time, but TNC is far better funded than most land trusts and has a more explicit aim of employing strategic, science-based approaches to conservation. It simply is not known whether land trusts overall provide an efficient mechanism for conservation of natural capital either in terms of the benefits provided or in terms of efficiency (Are payments being made for land that would have remained undeveloped even without payments?).

On the social side, there is ongoing debate as to whether conservation easements represent a net benefit to the public, given their impact on tax revenue and the (generally unquantified) public benefits provided by lands. The net impact of conservation easements on tax revenue in unclear. On the one hand, granting deductions for the reduced land value of conservation easement lands reduces overall tax revenues. Moreover, easements reduce property values for the property with the easement, and therefore reduce the taxes collected from the property. In this way, the general taxpayer is subsidizing conservation because of forgone tax revenues that have to be made up from other sources. On the other hand, property values have been shown to increase with proximity to open space, so conservation easements could increase taxes in a region.

Conservation easements may also affect the equity of the distribution of benefits across society, depending on where they are located and who has access to them. Although many ecosystem service benefits, such as clean water and scenic views, can be provided by easements regardless of whether the public can directly access that land, a lack of access does preclude other benefits, such as recreational opportunities. A 2009 survey of land trusts found that nearly 80 percent of responding land trusts provide public access to at least some of their lands, but that most land (59 percent) protected by conservation easements is not accessible to the public.

Conservation easements have the potential to augment or complement protected areas when located in buffer zones or corridors. In addition, the permanence of easements (with some exceptions) may make them more effective than government designation, which can reverse protection due to pressure from interest groups or a lack of funding. The flexibility of conservation easements and the ability of landowners to maintain property rights also make them appealing in many situations, and this approach has spread in Latin America and Europe as well.

Key Lessons Learned

Land trusts and the conservation easement model have been successful in the United States at enlisting privately owned land for conservation purposes and appears to be reducing development in these areas. Like so many other financial mechanisms for conservation covered in this book, conservation easements again highlight the challenges associated with monitoring the effectiveness of the mechanism, and in particular the trade-off between metrics that are easy to measure (area under easements) versus metrics that are more difficult to gather but more precisely reflect conservation values (ecosystem service benefits provided, biodiversity protected). The role of the government in encouraging conservation easements through the tax code is also notable. The tax breaks associated with easements mean that, in effect, the general taxpayer subsidizes conservation. Land trusts and private landowners together determine where and how to conserve (fig. 8.1), without direct involvement from the government. Some argue that the uncertain fiscal and equity implications of conservation easements for the government, along with the rarity of external verification, call for greater government oversight.

Conclusions

As illustrated by the examples in this chapter, voluntary mechanisms can contribute to conservation of natural capital on both privately owned and government-managed lands. Successful implementation of these mechanisms requires a match between the type and scale of benefits provided by a site and the motivations of the donors. The Project Finance for Permanence model, as illustrated by the Amazon Region Protected Areas Transition Trust, has successfully aggregated funding from global donors in support of securing the global benefits provided by Amazon forests, namely global climate regulation and global biodiversity values. Land trusts in the United States, on the other hand, often—though not always—operate more locally, and thus may focus more on local benefits, such as preserving aesthetic and cultural values, as much as biodiversity conservation. In both types of mechanisms, governments, nongovernmental organizations, and private actors all have a vital role to play in enabling their success.

Key References

Land Trust Alliance. 2016. *2015 National Land Trust Census Report*. http://s3.amazonaws .com/landtrustalliance.org/2015NationalLandTrustCensusReport.pdf.

Linden, Larry, Steve McCormick, Ivan Barkhorn, Roger Ullman, Guillermo Castilleja, Dan Winterson, and Lee Green. 2012. "A big deal for conservation." *Stanford Social Innovation Review* 43–49.

Merenlender, Adina M., Lynn Huntsinger, Greig Guthey, and S. K. Fairfax. 2004. "Land trusts and conservation easements: Who is conserving what for whom?" *Conservation Biology* 18, no. 1:65–76.

Owley, Jessica, and Adena R. Rissman. 2016. "Trends in private land conservation: Increasing complexity, shifting conservation purposes and allowable private land uses." *Land Use Policy* 51:76–84.

Soares-Filho, Britaldo. 2016. *Role of Amazon Protected Areas, Especially the Conservation Units Supported by ARPA, in Reducing Deforestation*. Rio de Janeiro: Funbio.

Soares-Filho, Britaldo, Paulo Moutinho, Daniel Nepstad, Anthony Anderson, Hermann Rodrigues, Ricardo Garcia, Laura Dietzsch et al. 2010. "Role of Brazilian Amazon protected areas in climate change mitigation." *Proceedings of the National Academy of Sciences* 107, no. 24:10821–26.

WWF. 2015. *Project Finance for Permanence: Key Outcomes and Lessons Learned*. Washington, DC: World Wildlife Fund. https://www.worldwildlife.org/publications/project-finance -for-permanence-key-outcomes-and-lessons-learned.

CHAPTER 9

Water Funds

Kate A. Brauman, Rebecca Benner, Silvia Benitez, Leah Bremer, and Kari Vigerstøl

Clean and abundant water for people, cities, and agriculture is threatened by upstream deforestation, poor agricultural practices, and development. Managing watersheds can help regulate downstream water flow and reduce pollutants in water. Recognizing this, investments in nature-based source water protection are becoming more common. However, these investments do not occur where communities lack funding or the institutions to direct funding from downstream beneficiaries to upstream residents. Here we describe water funds, a replicable financial and governance mechanism to link downstream water users to upstream residents, paying them to manage watersheds sustainably. Water funds include mechanisms that allow multiple stakeholders to provide (1) funding, to collect resources for watershed management over the long term; (2) governance, to enable transparent and inclusive joint planning and decision making; and (3) watershed management, to plan, implement, and monitor activities on the ground. Water funds must be tailored to the sociocultural, economic, and ecological context of their location, so each takes its own specific form, but all have all three elements. Their success depends both on the pooled financial resources from downstream supporters and on the coordinated support of upstream residents, who participate because they meaningfully benefit from water fund activities. The model has spread rapidly through Latin America and the world.

Demand for clean, abundant water is increasing, yet the supply of clean water is increasingly at risk. As of 2010, about 40 percent of watersheds upstream of large urban areas had high to moderate levels of land and vegetation degradation caused by deforestation, poor agricultural practices, and development.

This can reduce the amount of available clean water or increase water treatment costs downstream.

While built infrastructure to move and treat water will always be necessary, nature-based solutions can improve water quantity and quality, reducing costs and increasing reliability. Managing upstream land cover and land use can help regulate water flow and reduce pollutants in water, improving human well-being downstream while protecting terrestrial and aquatic ecosystems.

Despite these advantages, people often fail to capture the benefits of nature-based source water protection. There are a variety of reasons for this, including, for example, (1) spatial displacement between those in water source areas and those who use water; (2) insufficient knowledge about the specific water security benefits of watershed management; (3) lack of funds by those who could implement watershed management strategies; and (4) limited mechanisms for funding to move from beneficiaries to suppliers.

Despite these challenges, nature-based investments to secure clean water are becoming increasingly common. A survey of the state of watershed investment found that, in 2015, payments to conserve or restore watersheds totaled about US$24.6 billion worldwide. This encompassed 419 programs in sixty-two countries affecting at least 487 million hectares (an area larger than India). Most of the payments that arose through these schemes went to private landholders, in total nearly US$10 billion in 2015.

What Is a Water Fund?

Water funds are one type of investment in watershed services. The label *water fund* refers to a replicable financial and governance model developed by cities, development banks (such as the Inter-American Development Bank), and conservation practitioners (such as The Nature Conservancy). This funding and governance mechanism links downstream water users to upstream residents, compensating them for managing watersheds sustainably. Water funds can create a virtuous cycle so that fund investments, when well designed and equitable, create opportunities and support for upstream residents to manage their land in ways that improve water resources for downstream water users, who provide continued political and economic support, as well as benefiting upstream communities.

Water funds both mobilize funding and coordinate watershed management activities. Water funds have been referred to as "collective-action funds"

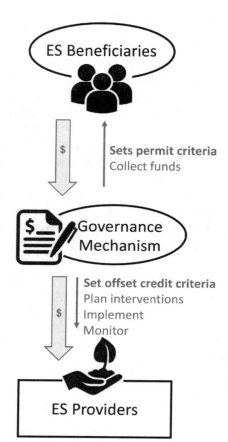

Figure 9.1. Water funds connect ecosystem service providers and ecosystem service beneficiaries through an interconnected governance mechanism, funding mechanism, and management mechanism.

because they are characterized by pooled resources and coordinated action across a landscape. The success of a water fund depends on both financial resources and engagement of upstream residents, who participate because they benefit from water fund activities in a meaningful way (fig. 9.1).

Water funds have three primary organizational components:

- **a funding mechanism** to collect and provide resources for watershed management in the long term;
- **a governance mechanism** for joint planning and decision making; and

- **a watershed management mechanism** to carry out funded conservation and management activities.

The funding mechanism allows multiple stakeholders—including water users (public and private), government agencies, development banks, and non-governmental organizations (NGOs)—to coordinate and provide long-term resources for source water protection.

The governance mechanism is a board or a project management unit composed of a diverse set of stakeholders, potentially including representatives of investors, the public sector, water companies, and local communities both upstream and downstream. The multi-stakeholder structure of this group, as well as transparency in how members are chosen and how decisions are made, builds trust and engagement among watershed stakeholders for project planning and decision making.

The watershed management mechanism is a group that plans, targets, implements, and monitors activities on the ground. Each water fund has unique objectives and goals but, in general, they work to (1) improve or maintain water quality and water quantity for downstream users; (2) maintain regular flows of water throughout the year; (3) maintain or enhance biodiversity, both freshwater and terrestrial; and (4) improve or maintain human well-being for upstream communities. Management mechanisms often seek to adopt a science-based approach to improve the impact and cost-effectiveness of watershed interventions.

History of Water Funds

There are many approaches to source water protection. For example, some cities in the United States have purchased land in their source watersheds to ensure that polluting activities are not undertaken there. Chapter 6 discusses New York City, which invested in watershed protection to avoid building a costly water filtration plant. Other payment models exist as well. For example, in 2003 Fundación Natura initiated the "Watershared" model in Bolivia when downstream irrigators negotiated to provide alternative development tools like beehives or fruit trees in exchange for upstream landowners conserving forests by fencing out cattle.

The term water fund refers to a particular set of watershed conservation approaches that include governance, financial, and land management mechanisms. The water fund idea was conceived in the late 1990s in Quito, Ecuador, a growing city with substantial water demands. There was concern that land use in Quito's watershed could compromise the water supply. However, many farmers and ranchers depended on the watershed for their livelihoods, making management difficult. The possibility arose of water users making voluntary financial contributions to compensate farmers and ranchers for watershed management. In 1997, The Nature Conservancy (TNC) and its partners began negotiations with the municipality of Quito and Quito's water utility (now Empresa Metropolitana–EPMAPS). The fund launched in 2000 with voluntary donations from EPMAPS and TNC going into a trust fund that formed the financial basis of the water fund. Several other water users and institutions subsequently joined the water fund.

Following the creation of the water fund in Quito, the model has spread rapidly through Latin America and the world. In 2011, TNC joined the Inter-American Development Bank, FEMSA Foundation, and the Global Environmental Facility to support existing water funds and promote the creation of new water funds in Latin America and the Caribbean. The Latin American Water Fund Partnership initiative has been key for the growth and support of water funds in the region, with twenty funds in operation and approximately twenty more in design. TNC and its partners have moved the idea of water funds to multiple countries and continents around the globe. As of 2017, TNC has a portfolio of twenty-nine funds in operation and approximately thirty in design (fig. 9.2). TNC's goal is to improve the water fund model to reduce the risks associated with source water-protection investments.

Variation among Water Funds

Within the structure of a water fund, each takes its own form and function because it must be tailored to the sociocultural, ecological, and economic context of its location. As a result, water funds display a diversity of funding, governance, and management strategies, though they all organize and mobilize resources and implement watershed management. Table 9.1 specifies key elements of water funds and includes details about how those differ among funds.

Figure 9.2. Location of water funds around the world. These include both operational and pre-operational water funds tracked by The Nature Conservancy as of March 2018. Solid diamonds correspond to water funds highlighted in this chapter. Data courtesy of Emily Simmons, The Nature Conservancy.

Three Case Studies

Three case studies presented here illustrate similarities and diversity among operational water funds.

The Fund for the Protection of Water (FONAG) in Quito, Ecuador, is the oldest official water fund. FONAG has an endowment of more than US$11 million, a board of directors as the decision-making body, and a technical secretariat that implements watershed management activities.

The Upper Tana-Nairobi Water Fund, officially launched in 2015, works in the Tana River watershed in Kenya. This watershed supplies 95 percent of Nairobi's water, 50 percent of Kenya's electricity, and is home to over a million farming families. The fund was set up with multiple objectives, including improving agricultural livelihoods upstream and securing Nairobi's water supplies by reducing erosion and maintaining dry season flow.

The Camboriú Water Producer Project serves the municipality of Balneário Camboriú, Brazil, an important tourist destination. The program aims to reduce sediment at the municipal drinking water intake by paying landowners to conserve and restore forests along rivers and in steeply sloped areas.

Table 9.1. Diversity across Water Funds

City or state of Water Fund	Objectives	Primary funding source(s)	Role of government	Participants (who carries out protection activities)	Program activities
Cauca Valley, Colombia	Water quality (sediment) and quantity; biodiversity conservation; livelihood improvements	Private companies; government; environmental agencies; private landholders	State funding	Indigenous communities; rural landholders	Forest protection; revegetation; agroforestry; livestock management; education; strengthening local organizations
Medellin, Colombia	Water quality (sediment and nutrients)	Water company; government environmental agencies; private companies; municipality; NGOs	Municipal and regional government funding; role on governance board; alignment with public policies and agencies	Private and communal landowners supported by water fund implementation team. Contractors hired for specific work (e.g., planting trees, studies)	Targeted land protection; restoration; improved agricultural practices; education; sanitation in rural areas
Quito, Ecuador	Water quality and quantity	Water company; electric company; NGO; private companies; multi-laterals; bilaterals; municipality	Public utility funding; role on governance board; alignment with public policies and agencies	Private and communal landowners supported by Water Fund implementation team. Contractors for specific work	Targeted land protection; restoration; improved agricultural practices; land acquisition; education; water management
Lima, Peru	Sustainable water management	Private companies; NGOs; government environmental agencies	No major role currently; federal water regulator now requires all water companies to invest in natural infrastructure	Private and communal landowners supported by Water Fund implementation team. Contractors for specific work	Targeted land protection; restoration; water treatment; education; improved agricultural practices; governance; traditional water technology

City or state of water fund	Objectives	Primary funding source(s)	Role of government	Participants (who carries out protection activities)	Program activities
São Paulo, Brazil	Water quantity (flow regulation) and quality (sediment); environmental awareness; rural incomes	Municipalities; watershed committees; government water agency; state environmental agency; NGOs	Primarily publicly funded and run; municipal tax and state funding	Farmers in Extrema and other municipalities in source watersheds	Restoration; forest protection; soil conservation; dirt road management
Rio de Janeiro, Brazil	Water quantity (flow regulation) and quality (sediment, contamination); biodiversity conservation; rural incomes	Watershed committee; state environmental agency; NGO	Primarily publicly funded and run; municipal tax and state funding	Farmers in Rio Claro and other upstream municipalities	Restoration; forest protection; soil conservation; dirt road management; rural wastewater services
Camboriú, Brazil	Water quantity (tourist season) and quality (sediment)	Municipality; water company; NGO	Primarily publicly funded and run; public utility and state funding	Farmers in Camboriú municipality	Forest protection; dirt road management; restoration
Nairobi, Kenya	Water quality (sediment); engagement and participatory land and water planning	Sewer and water company; electric company; government agencies; private companies; NGOs	Role on governance board; engagement and technical expertise.	NGOs, farmer associations, civil society organizations working with land stewards and small-scale farmers	Improved agricultural practices; reforestation
New Mexico, USA	Water quality (postfire soil and debris); forest restoration	Federal, state, and local government agencies; water utilities; private companies; individuals	State funding; role on governance board; role in implementation	Federal government; local logging companies; NGOs	Forest thinning; restoration; education

Case 1. The Fund for the Protection of Water

The Problem

Beginning in the 1990s, there was concern that upstream land use by numerous and diverse farmers and ranchers could compromise the supply of water for Quito, Ecuador. The possibility arose of compensating farmers and ranchers for watershed management. The Fund for the Protection of Water (FONAG), created in 2000, is a private fiduciary fund that manages both public and private watershed areas and had an endowment of more than US$12 million in July 2017. It has a board of directors as the governance body made up of the Quito Water Company, the Quito Electric Company, TNC, a privately owned brewery, a bottler, and a local NGO. The management mechanism is a technical secretariat.

In addition to direct source watershed-protection activities, FONAG has worked with over 400 local families to strengthen watershed governance, environmental education, and communication. FONAG's vision is to mobilize all actors in the watershed to exercise their civic responsibility on behalf of nature, particularly water resources.

The Ecosystem Services

The watersheds that provide water for Quito are located high in the Andes. Much of this area is *páramo*, a high-altitude shrub and grassland ecosystem; the watershed also includes high-elevation wetlands and forest. Because of their deep organic soils, *páramos* are crucial for water regulation. Activities such as grazing (e.g., cattle, sheep); agriculture (e.g., potatoes); fire used in agricultural management; and afforestation, which can compact soils, alter water use regimes and degrade water quality (e.g., sediments, turbidity, nutrients, bacteria). FONAG has prioritized the location of its work based on development threats and on the importance of individual sub-watersheds to the municipal water supply.

In addition to their importance for water, the ecosystems FONAG manages are important for biodiversity and carbon storage. *Páramo* is characterized by a high percentage of endemic species. Threatened species including the Andean bear, the Andean tapir, and the Andean condor are found there. The *Polylepis* forest is a species endemic to the Andes, and Andean wetlands have been found to store very high values of carbon.

Ecosystem Service Beneficiaries

FONAG initially had two main members: TNC and the Quito Water Company (the key water user). Other water users have since joined: the Quito Electric Company in 2001; private organizations including a beer company in 2003; the Swiss Agency for Development and Cooperation in 2005 (which ceded its space on the board to a local NGO); and a water bottling company in 2005. The main incentive for the water utility and the other major water users is avoided or reduced future costs for water treatment and supply. For TNC, the incentive is long-term financing for conserving nature.

Ecosystem Service Suppliers

There are a wide variety of suppliers in Quito's water supply area. Some of the source watersheds are national protected areas managed by the Ministry of the Environment. In addition, both the Quito water company and FONAG itself own land in the watershed. The remaining suppliers are a mix of small-scale subsistence ranchers and farmers; communal landowners that own and manage land under communal rules (these are often indigenous communities); and large private landowners.

Terms of the Exchange: Quid Pro Quo

FONAG runs a variety of programs to conserve and restore watersheds areas on lands owned by the water company, by the fund, by private landowners, and by well-organized communities.

Management of Land Owned by the Fund and Its Partners

FONAG manages more than 16,000 hectares of land owned by the Quito Water Company and by FONAG. To ensure protection of *páramos* and wetlands, the fund invests in fencing and hiring local people as "*páramo* guards" to help control threats such as cattle and fire. It has been common for decades to raise cattle and sheep in the *páramo*, in riparian areas, and in wetlands, so restoration is necessary in some areas. Restoration may be passive (e.g., fencing to exclude cattle and allow the *páramo* to restore itself) or active (e.g., replanting native species).

When FONAG was created in 2000, the national protected areas located in Quito's source watershed were underfunded. FONAG provided important support for the management of those areas by implementing key activities such as park guards. As the Ministry of Environment gained capacity to manage and fund their protected areas, FONAG began targeting its work outside those boundaries, though it still maintains a high level of coordination and collaboration with the Ministry.

Work with Communal and Private Landowners
Since the beginning, FONAG has worked with local communities and farmers on conservation and restoration. In most cases, landowners set aside key areas and implement conservation activities; FONAG supports the landowner with an incentive or in-kind payment. Incentives include:

- supplies for fencing
- technical and in-kind help to plant live fences
- seeds for revegetation
- technical assistance to improve farm activities
- support for alternative income sources like guinea pig farms
- support for establishment of alternative food sources like organic vegetable gardens
- support for a community to switch from raising cattle to raising alpacas (Andean camelids, which have valuable wool)
- fire control training

In addition to direct management to protect and restore natural ecosystems, FONAG has programs to promote good watershed governance and sustainable watershed management. These include the following:

- Environmental education: More than 45,000 people have been involved in a program teaching sustainable watershed management.
- Water management: To improve the quality and accessibility of watershed data for decision making, including close collaboration with Ecuador's National Water Agency to improve and update information on water concessions.
- Communication: Promoting the importance of sustainable watershed management and the work and results of the water fund.

Mechanism for Transfer of Value
FONAG has an average annual budget of US$2 million, which comes from interest from the endowment, annual contributions from board members (30

percent of board members' annual contribution goes to the annual budget of the fund, the other 70 percent is invested in the endowment), and additional donations or contributions.

FONAG's endowment now yields about US$400,000 in interest each year. In 2000, FONAG had US$21,000 in its endowment; as of 2017 the endowment was more than US$12 million.

The public members of FONAG's board provide the largest source of funding: the Quito Water Company provides more than 80 percent of annual contributions and the Quito Electric Company provides the next largest contribution. Decision-making power on the board is linked to these monetary contributions. In 2006, FONAG, with the support of its board members, helped to pass a municipal bylaw requiring the Quito Water Company to provide 2 percent of its revenue to the water fund. This was an increase from the initial 1 percent voluntary commitment. This has been essential for ensuring FONAG's long-term financial security.

FONAG has also received funding from development banks (e.g., the Inter-American Development Bank) and bilateral cooperation. The United States Agency for International Development (USAID) was an important donor that supported and strengthened FONAG during its early years. Grants and donations have come from other organizations interested in supporting the objectives of the Water Fund, including corporate donors such as Coca-Cola and its local bottler.

To implement its programs, FONAG signs conservation agreements with landowners that set the conditions for collaborative work. The technical team of FONAG negotiates with individual landowners to determine the type of incentive appropriate to local needs and interests. All work is paid from the annual budget of FONAG, and FONAG's staff of about forty people ensure that communities receive their in-kind payments or incentives. Other employees of the management unit include a director (technical secretariat), technical director, community park guards, hydrologists in charge of monitoring, a technical team for restoration activities, communication staff, administrative staff, and a team that leads the education work at local schools.

Monitoring and Verification

FONAG has recently undertaken a number of monitoring and evaluation projects to measure the impacts of its work. It is monitoring water flows and water quality in several key watersheds and using this information to improve the effectiveness of investments, particularly with respect to water company objectives, and for communication to potential investors. FONAG is also measuring terrestrial and freshwater ecosystem integrity. In addition, it has established partnerships with universities for research and monitoring.

Effectiveness

FONAG has successfully protected and restored over 25,000 hectares of *páramo*, wetland, and Andean forest. Early monitoring work and academic studies show evidence of systematic positive impacts on water quality and ecosystem integrity in the watersheds where FONAG is working. There have also been reductions in grazing impacts and fire damage.

With support of the Latin American Water Funds Partnership, FONAG is implementing a return-on-investment study to evaluate the economic benefits to the Quito Water Company from investing in watershed conservation. Initial analysis suggests a positive return on investment for FONAG´s interventions over the next twenty years. This results from financial savings for water treatment for the Quito Water Company because of a reduction in nutrients, bacteria, turbidity, and sediments in source water, as well as increased water flows.

Case 2. Upper Tana-Nairobi Water Fund

The Problem

The Uppe Tanar-Nairobi Water Fund was created to help provide clean, reliable water to the city of Nairobi. Over 90 percent of Nairobi's water comes from the Upper Tana River. The Nairobi City Water and Sewerage Company was facing declining water quality, especially during storm events, because of sediment in the water supply. In addition, the country's largest hydropower company, KenGen, was affected by sediment in its hydropower reservoirs on the Upper Tana River. KenGen's hydropower provides half of Kenya's electricity.

Exploration of a water fund for the basin began in 2012 with six founding partners, although analysis and discussion of source water protection as a viable solution had occurred previously. After several years of technical analyses, partnership building, development of a governance structure, prioritization of activities to meet desired outcomes, and implementation of pilot projects in the basin, the Fund officially launched in March 2015.

The Nairobi Water Fund is an independent charitable trust with a governance board consisting of the county government, the agricultural minister, the council of governors, the Nairobi water company, a beverage company, and a hydropower company.

The Ecosystem Services

The headwaters of the Tana River lie in the Aberdare National Park, home to rare and endemic wildlife, including the endangered mountain bongo antelope and colombus monkeys. On its 1,000 kilometers run to the Indian Ocean, the river passes through forests, small-scale agricultural communities, and small towns before reaching the drinking water intake for the city of Nairobi. Since the 1970s, the watershed has undergone a huge expansion of farming, mostly tea, coffee, bananas, and other crops farmed by small farmers. Land scarcity and declining soil productivity due to erosion have driven farmers to expand cultivation onto steep slopes and into riparian areas. This has led to a loss of groundcover, leaving bare slopes vulnerable to erosion. During the rainy season, massive amounts of sediment wash into the river.

The water fund aims to strengthen the delivery of reliable, clean water and reduce the impacts of sediment by restoring degraded and agricultural lands to reduce erosion. Activities include the following:

- Planting native tree species in high-altitude areas of the watershed that have been cleared for agriculture but are not currently productive farmland
- Planting buffer zones of trees and plants along rivers and streams to slow the conveyance of runoff
- Stabilizing soil by terracing crop fields
- Applying mulch to secure soil from washout and to restore soil fertility
- Growing buffer zones to retain soil and ease water conveyance, such as planting riparian habitat and bamboo to stabilize banks
- Creating water trapping pits at the top of the catchment to store and capture excess rainwater
- Reducing water withdrawals from the river by installing rain tanks and drip irrigation to reduce irrigation pressure during dry periods

Ecosystem Service Beneficiaries

The primary beneficiaries of the Upper Tana-Nairobi Water Fund are the residents of Nairobi, Nairobi City Water and Sewerage Company, and the hydropower generating company KenGen. The river supplies drinking water for 4 million residents of Nairobi and an additional 5 million people in the watershed. In addition, there are five hydropower dams on the main stem of the Tana River with an installed capacity of 543 MW.

Nairobi's population has doubled in the last twenty-five years and will continue to rise, demanding increasing supplies of food, water, and electricity from the Upper Tana River basin. Yet the heavy sediment load increases water treatment costs by more than 33 percent as sediment spikes during the wet season. Reservoirs lose active storage capacity as they fill with sediment, limiting the ability of KenGen to balance production across seasons. New agricultural lands have increased demand for irrigation water, competing with the water needs of a city that already has a 30 percent deficit. The Fund could help alleviate all of these issues.

There are many additional beneficiaries of this water fund, including farmers who are engaged in the Fund via increased agricultural productivity, cities along the Tana River who also draw their water from the river, the people of Kenya who depend on hydropower for their electricity, and everyone who benefits from ecosystems that are protected or restored across the basin. Expected benefits to a range of stakeholders are listed in table 9.2.

Ecosystem Service Suppliers

There are 300,000 small farms on the steep slopes in the Upper Tana watershed. About 98 percent of the people who live in the watershed are farmers who grow tea, coffee, bananas, and other crops. It is estimated that about 50,000 of these farmers are located in the steepest and most critical areas.

Terms of the Exchange: Quid Pro Quo

Similar to FONAG, the Upper Tana-Nairobi Water Fund uses in-kind compensation mechanisms to encourage farmers to adopt agricultural best management practices, install riparian buffers, and reforest. These in-kind compensation packages include provision of water pans, seeds, equipment, and livestock. Compensation also includes capacity building and training around agricultural production. Farmers enrolled in the water fund also act as conservation advocates throughout the community, spreading the word about conservation through community meetings and local social groups.

Table 9.2. Expected Benefits of the Upper Tana-Nairobi Water Fund

Stakeholder	Benefit
Nairobi City Water and Sewerage Company (NCWSC)	- Municipal water users - Reduction in wet sludge disposal and treatment costs - Increased dry season flows - Greater supply of water
Kenya Electricity Generation Company (KenGen)	- Reduction in reservoir sedimentation - Avoided turbine intake maintenance costs
Upstream farmers	- Increased soil and water conservation and fodder for livestock - Additional income and employment opportunities
Urban private sector processors	- Improved water supply
Local communities	- Cleaner drinking water
General: Ecosystem services	- More habitat for pollinators - Increased carbon storage in new trees planted

Source: Adapted from Vogl et al. 2016.

Note: Anticipated benefits of source water protection in the Upper Tana River basin and recipient stakeholder groups.

Mechanism for Transfer of Value

Similar to FONAG, the Upper Tana-Nairobi Water Fund is a private-public partnership supported by a combination of voluntary contributions and multilateral funders. In the first four years, the water fund was able to mobilize US$4 million through voluntary contributions. Board members are not required to provide an annual contribution, though many do contribute to the fund. For example, the Nairobi City Water and Sewerage Company imposed a 0.5 percent tariff on water fees to support the water fund and is an important contributor. There are also important multilateral funders, including the Global Environment Facility (GEF), which committed US$1 million in seed funding. The project aims for a US$15 million endowment that will provide funding over the long term.

To implement its work, the water fund works with farmers using the social capital built by nonprofit and for-profit agricultural and tree-planting organizations and agencies that have worked in the upper watershed over many years. These include the Green Belt Movement, SACDEP, Water Resource Users Association, Kenya National Farmers Federation, and other civil society organizations.

Monitoring and Verification

The Upper Tana-Nairobi Water Fund has developed a robust monitoring and evaluation plan measuring both the scale of action and the delivery of a wide range of desired outcomes, much of which has been implemented. The plan involves monitoring (1) how the water fund is being institutionalized, measured by indicators, including the amount of funding in the endowment, and the number of meetings held by the Steering Committee and Board of Trustees; (2) the water fund's support of livelihoods, food security, and economic development, measured by indicators including the number of people receiving project services, the land area involved, and the number of drip irrigation systems installed; and (3) the status of knowledge management and learning systems, measured by indicators including the number of water monitoring stations installed, the number of schools participating in the awareness program, and the number of government policies to protect soil and water. The water fund will also collect data and conduct analysis including these outcome indicators:

- Turbidity at multiple basins within the watershed
- Sediment load flowing into reservoirs
- Number of days per year with maximum turbidity at water intake low enough to avoid switching to more expensive water clarifiers
- Decrease in river water abstracted by smallholder farming households after installing drip irrigation and/or a rainwater pan
- Increase in crop productivity after installing drip irrigation and/or a rainwater pan
- Increase in households with improved Multidimensional Poverty Assessment Tool score
- Increase in households saying permanent vegetation cover has increased and that soil erosion has decreased on their farm
- Project lessons reflected in government policies, strategies, or programs

Effectiveness

The water fund has engaged over 10,000 farmers, helped bring 120,000 acres of farmland under sustainable management, and planted 175,000 trees annually.

Table 9.3. Expected monetary benefits to major beneficiaries of the Upper Tana-Nairobi Water Fund

Stakeholder	Benefit or Cost	Present Value (US$)
Water Fund	Cost: Investment cost (voluntary contributions from Board and multilateral investors)	(7,110,000)
Agricultural producers in watershed	Cost: Maintenance of erosion control structures, etc. Benefit: Increased agricultural productivity	(8,520,000) 12,000,000
Nairobi City Water and Sewerage Company (NCWSC)	Benefit: Avoided water treatment cost (flocculants and electricity), revenue from saved process water, future savings assuming increased future demand	3,390,000
Kenya Electricity Generation Company (KenGen)	Benefit: Avoided interruptions, increased generation from increased water yield	6,150,000
	Present value of benefits	**21,500,000**
	Present value of costs	**(15,600,000)**
	Net present value	**5,900,000**

Note: Predicted benefits are over a thirty-year time frame. Adapted from Vogl et al. 2016.

Farmer participation has approximately doubled each year since engagement began in 2013.

A business case study was conducted on the Upper Tana-Nairobi Water Fund that modeled the projected cost savings and other benefits of implementing the water fund at scale. At full implementation, some of the expected outcomes include over 50 percent reduction in sediment concentration in rivers (varying by watershed and time of year); 18 percent decrease in annual sedimentation in a key reservoir; and improved water quality, with a potential decrease in waterborne pathogens, for more than half a million people. Expected monetary savings are detailed in table 9.3. In total, a US$10 million investment in water fund interventions is expected to return US$21.5 million in economic benefits over a thirty-year period.

Case 3. *Camboriú*

The Problem

The coastal municipality of Balneário Camboriú is an important tourist destination in Brazil. During the summer tourism season, the permanent population of 170,000 people increases to more than 800,000. Balneário Camboriú has year-round concerns about elevated sediment and additional concerns about water quantity during the tourist season, when demand more than quadruples. The Balneário Camboriú Water Company treats water for both the Balneário Camboriú municipality and the Camboriú municipality, inland of Balneário Camboriú.

The Camboriú Water Fund is an initiative of the Balneário Camboriú Water Company and partners including TNC, the National Water Agency, and the municipalities of Camboriú and Balneário Camboriú. Funding for the water fund comes from a municipal law predating the Water Fund that requires that 1 percent of the water company's budget goes to finance watershed management initiatives. After visiting Brazil's first water producer project in Extrema, São Paulo, an innovative local water company employee took advantage of the opportunity to finance a water producer program using these legally mandated funds.

The Camboriú Water Producer Project has inspired TNC to work with the state water regulator to allow the local water company and other Brazilian water companies to charge a water tariff for conservation. They are working to scale this approach with other water regulators and water companies throughout Brazil.

The Ecosystem Services

Source waters for Camboriú and Balneário Camboriú lie in the Atlantic Forest Mountains. While forest cover remains high in the immediate regions surrounding Camboriú, the Atlantic Forest is the most threatened forest biome in Brazil, and there is concern about future deforestation for agricultural and urban expansion.

The water company faces high levels of sediment in source water, which increase treatment costs, as well as frequent water shortages during the tourist season. Based on the assumption that deforestation is a major cause of sediment, the water fund aims to conserve and restore the forest. Following Brazil's Forest Code, the water fund prioritizes reforestation in riparian buffers and on steeply sloped areas. In Camboriú, there is high forest cover and thus a seedbank available, so the program is considering passive restoration in areas needing restoration. There is also a focus on improving dirt roads, which are a major source of sediment.

Another major challenge facing the region is increasing urbanization in lower elevation areas that are currently used for rice cultivation. The program is considering strategies to address this, but it is currently beyond the program scope.

Ecosystem Service Beneficiaries

The program aims to reduce total suspended solids at the municipal drinking water intake, thereby reducing water losses and treatment costs. The primary beneficiary is the Balneário Camboriú Water Company, which provides water for tourists, residents of Balneário Camboriú, and the upland municipality of Camboriú.

Ecosystem Service Suppliers

Land in the source watershed is owned primarily by farmers with large to medium-sized landholdings. Program managers have worked hard to build trust in a region where recruitment is challenging given a long history of distrust of government agencies. The program has also faced the challenge of building trust between the municipalities of Balneário Camboriú and Camboriú, which share the water system. Because of the benefits of tourism, Balneário Camboriú is one of the five wealthiest municipalities in Brazil. Camboriú, the inland municipality where the water suppliers live, is ranked much lower. Despite providing water to the Camboriú Water Company, the Balneário Camboriú Water Company still needed to invest in trust building in order to work in the Camboriú watershed. Though anyone with de facto land tenure can enroll in the water fund, it has been challenging to get people to enroll; there is currently more money available than can be spent.

Terms of the Exchange: Quid Pro Quo

Unlike FONAG and the Upper Tana-Nairobi Water Fund, in Camboriú farmers receive direct compensation of 300 Brazilian reals per year (~US$100), paid twice per year, for restoration and conservation. Contracts primarily focus on forest conservation, but unlike other areas in Brazil where participants may receive more money for conservation than for restoration, in Camboriú all areas enrolled in the program are compensated equally.

Critical to the context in this region is that Brazil's Forest Code mandates conservation and restoration of streamside and hilltop priority zones. Thus, the program provides a benefit to participants by helping them comply with the law. However, the Camboriú area has had consistently poor enforcement of the Forest Code, so it is unclear if this provides motivation.

Mechanism for Transfer of Value

The Balneário Camboriú Water Company is the most important funder in the project, contributing from a pool of money set aside by a municipal law that requires that 1 percent of the Water Company's budget goes to finance watershed management initiatives.

The water company is the primary project manager, although other members of the project management unit participate in monitoring, site verification, and outreach.

Monitoring and Verification

Camboriú has partnered with the state meteorological organization to monitor sediment and flow. They use a before-after-control-impact design. Verification of properties and activities is done once every six months with the water company and at least one other member of the project management unit.

Effectiveness

Camboriú is a new program, so it is difficult to quantify the ecosystem service outcomes of the program. As of 2017, seventeen landowners are currently enrolled, with approximately 20 hectares in restoration and 300 hectares in conservation.

Hydrologic and economic models of the project have been run to project likely effectiveness when the project is mature. Hydrologic models indicate substantial improvements in water quality and quantity from the program when it is fully enrolled. The costs and benefits of the Camboriú Water Fund over a thirty-year period (2015–2045) were not projected to be as large as the costs when only focusing on reducing sediment. Reductions in treatment costs and water losses could offset 80 percent of the water company's investment and 60 percent of the water fund's total cost. However, significant cobenefits, including reduced risk of flooding, were identified, increasing the water fund's overall benefits.

Conclusions

Water funds take a range of specific forms, but they are marked by integrating two distinct activities: mobilizing funding and coordinating watershed management activities. These actions are structured in three parts: a funding mechanism, a watershed management mechanism, and a mechanism to govern both.

Acquiring financial resources is far from the sole mark of success for a water fund; successfully implementing programs requires engagement by upstream residents. Building trust is critical. Many water funds have been developed in conjunction with existing local partners, leveraging their connections and social capital.

For upstream residents, meaningful benefits may include receipt of direct payments, but in many cases, water funds transfer value in the form of in-kind or technical support. In addition, environmental education programs are often a key part of watershed management activities, allowing the water fund to interface with more people and to ensure that residents understand the benefits their land management provides both downstream and locally. For downstream beneficiaries, engagement may require rigorous analysis demonstrating a return on investment, but in other cases, leadership and vision by a motivated and influential individual was key to implementation.

The expansion of water funds worldwide demonstrates their appeal. Biophysical and socioeconomic monitoring of outcomes are in their infancy, but continued investment to demonstrate the effectiveness of these investments will be critical.

Key References

Abell, Robin, Nigel Asquith, Giulio Boccaletti, Leah Bremer, Emily Chapin, Andrea Erickson-Quiroz, Jonathan Higgins et al. 2017. *Beyond the Source: The Environmental, Economic and Community Benefits of Source Water Protection.* Arlington, VA: The Nature Conservancy.

Bennett, Genevieve, and Franziska Ruef. 2016. *Alliances for Green Infrastructure: State of Watershed Investment 2016.* Washington, DC: Forest Trends Ecosystem Marketplace.

Brauman, Kate A., Gretchen C. Daily, T. Kaʾeo Duarte, and Harold A. Mooney. 2007. "The nature and value of ecosystem services: An overview highlighting hydrologic services." *Annual Review of Environment and Resources* 32:67–98. https://doi.org/10.1146/annurev.energy.32.031306.102758.

Bremer, Leah, Adrian L. Vogl, Bert De Bièvre, and Paulo Petry. 2016. *Bridging Theory and Practice for Hydrological Monitoring in Water Funds.* Latin America Water Funds Partnership.

Kroeger, Timm, Claudio Klemz, Daniel Shemie, Timothy Boucher, Jonathan R. B. Fisher, Eileen Acosta, P. James Dennedy-Frank et al. 2017. *Assessing the Return on Investment in Watershed Conservation: Best Practices Approach and Case Study for the Rio Camboriú PWS Program, Santa Catarina, Brazil.* Arlington, VA: The Nature Conservancy.

Vogl, Adrian L., Benjamin P. Bryant, Johannes E. Hunink, Stacie Wolny, Colin Apse, and Peter Droogers. 2016. "Valuing investments in sustainable land management in the Upper Tana River Basin, Kenya." *Journal of Environmental Management* 195 (1):78–91. https://doi.org/10.1016/j.jenvman.2016.10.013.

CHAPTER 10

Market-Based Mechanisms

Rebecca Chaplin-Kramer, Lisa Mandle, and Lauren Ferstandig

Market-based mechanisms have garnered attention for their potential to generate win-wins for natural capital and private enterprises. Through these mechanisms, businesses take on the costs of securing or enhancing natural capital, supported by demand from consumers of their goods or services. This chapter introduces two types of market-based mechanisms: eco-certification and impact investing. In the case of eco-certification, companies—motivated by brand differentiation, consumer loyalty, or a price premium for their products—contribute to maintaining or enhancing ecosystem services, often with their efforts or outcomes certified by third parties. We focus on ecotourism and supply-chain certification as two widespread examples of eco-certification. In the case of impact investing, consumers of a financial product pay into an investment vehicle that is designed to generate both financial returns and ecosystem service benefits. We illustrate the mechanism of impact investing with a private-equity example from the Murray-Darling Basin Water Sharing Investment Partnership in Australia, and a debt-based example from the DC Water Bond for nature-based stormwater infrastructure in the United States. With a number of long-established and widely adopted programs, the effectiveness of supply-chain certification has been well studied relative to other mechanisms and has generally produced positive environmental outcomes. Most examples of impact investing for ecosystem services are quite new and therefore have not yet been assessed. Both types of mechanisms hold promise for expanding funding directed at securing and enhancing natural capital beyond government and philanthropic sources.

....

Market-based transactions, in which consumers pay for the costs of securing or enhancing ecosystem services, are a mechanism for conservation finance that holds broad appeal. This appeal is based on the potential to generate win-wins for conservation and private enterprises, and to secure a greater pool of funding for preserving or enhancing natural capital than would be available through government or philanthropic funding. Here we focus on four case studies representing two types of market-based mechanisms: eco-certification and impact investing for ecosystem services.

In the case of eco-certification, consumers preferentially purchase products of superior environmental quality, based on third-party endorsement through certification. This can provide either a higher price or greater or more consistent demand to the producers willing to adhere to certification principles. We examine this mechanism through two case studies, including place-based certification in the form of ecotourism, and supply-chain certification, with a particular focus on coffee.

In the case of impact investing for ecosystem services, consumers of a financial product pay into an investment vehicle in anticipation of both financial and environmental returns. We provide an overview of the impact investing mechanism, and then focus on two case studies: a private equity example with the case study of the Murray-Darling Basin Water Fund in Australia, and a debt-based example with the case study of an environmental impact bond to fund nature-based solutions to reducing stormwater flows in Washington, DC.

Certification

Certification is a class of market-based mechanisms that can finance conservation of ecosystem services. Certification is meant to reward producers that adopt a more just or sustainable means of production and to allow consumers to live out their social values through their purchases. Certification can pertain to social responsibility, brand quality, or environmental stewardship such as energy efficiency or sustainable production practices. When applied to ecosystem services, certification aims to ensure that the use of land or water in the production of a good or service maintains or enhances ecosystem services (fig. 10.1).

Most certification is known as "third-party," which means that a separate entity is responsible for evaluating whether a business meets the certification standard's criteria, and continually auditing to ensure that this performance is maintained. These third-party certifying bodies often provide technical or other forms of assistance to producers to facilitate their implementation of the certification standards.

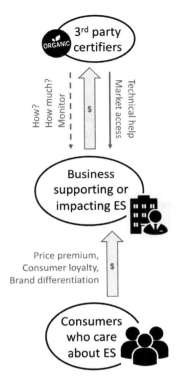

Figure 10.1. With eco-certification, consumers pay a higher price or are more likely to buy a product or service that is labeled sustainable, which then provides higher profitability or revenue to businesses for being more reliable ecosystem service providers. Third-party certifiers provide transparency and credibility to the consumer that the sustainability standards are being honored, and they may provide technical assistance or better market access to the producers.

Though not covered in our case studies, it is also worth noting that self-branding by companies that conserve natural capital or ecosystem services as part of their core business model or as a corporate initiative operates in a similar fashion. These individual commitments made by businesses may confer similar benefits in terms of consumer loyalty or willingness to pay, although without third-party certification.

Case 1. Place-Based Certification: Ecotourism

The Problem

In many parts of the world, rich biodiversity coincides with extreme poverty, and (over)harvesting or clearing natural habitat is often needed for people to meet even subsistence livelihood needs. If local people can receive payment from people wishing to enjoy that biodiversity through sustainable tourism (in a manner that respects the local culture and environment), there is a greater incentive for good stewardship rather than consumptive use of habitat. The potential to find win-wins for conservation and economic development goals garnered the support of many conservationists, national governments, and international aid agencies for ecotourism in the 1990s. However, the unintended impacts of "inappropriate" ecotourism on biodiversity has underscored the need to develop standards of practice and certification for ecotourism.

The Ecosystem Service

Tourism and recreation are the main focus for this mechanism, though, with the incentive provided by payment for these services, the improved stewardship of biodiversity and the ecosystems that attract the tourist visits could also support a number of other benefits, including other cultural and provisioning services, water, and climate regulation.

Ecosystem Service Beneficiaries

The main beneficiaries are the tourists, which could be a local, national, or international audience, depending on the scale of the attraction. But to the extent that there are secondary benefits supported by the ecotourism scheme, the beneficiaries may include local communities that gather cultural or subsistence products or derive other social benefit from the natural areas or rely on the area for drinking water, flood mitigation, or other services.

Ecosystem Service Suppliers

Tourism operators are obviously an important component of the delivery of this ecosystem service, but it is the users of the ecosystems to which tourists are drawn that are the suppliers of the biophysical supply of nature-based tourism. For ecotourism to be an effective scheme, the suppliers—the people who are otherwise incentivized to clear or degrade the habitat—need to be incentivized to maintain it in a healthier state that will continue to attract tourists.

Terms of the Exchange: Quid Pro Quo

While ecotourism is not a new concept, its certification is still somewhat nascent. The assumption of ecotourism enterprises is that by joining certification schemes, they will attract more patrons or at a higher price point than enterprises that are not certified. Another advantage of certification may be that certified operators gain preferential treatment from regulatory authorities, such as better access or longer operating licenses in parks and reserves.

In 1991, the Mohonk Agreement suggested a basic operational scheme and broad criteria for ecotourism certification. More recently, the Global Sustainable Tourism Council (GSTC) followed up with a set of more specific global criteria for certification and a plan to become the international authority for ecotour accreditation (i.e., certification for the certifiers), while allowing national or regional entities to assess how and to what extent each criterion should be met within their jurisdictions. GSTC specifies criteria and performance indicators that include conserving resources; reducing pollution; and conserving biodiversity, ecosystems, and landscapes.

While the GSTC standards are intended to encompass local or regional standards, they are too recent to evaluate. It is thus perhaps more instructive to look to local case studies of accreditation and operation to better understand how such programs work. The Costa Rican Certification for Sustainable Tourism (CST program) was the first performance-based voluntary environmental program created by a developing-country government, and hotels with higher environmental performance according to this certification have been able to establish price premiums (Rivera 2002). A more recent analog in the United States is Hawaii Eco-Tourism, which provides third-party certification of sustainable tourism across the state.

Mechanism for Transfer of Value

The mechanism of ecotourism and its certification can be best demonstrated through a profile of a particular business, and for this purpose Hawaii Forest and Trail (HFT) provides an excellent example. Certified by Hawaii Eco-Tourism, HFT licenses much of the lands on which they operate from private landholders, and in so doing raises awareness in both landholders and tourists of the value of maintaining the biotic integrity of such lands. HFT pays a gross percentage of their revenue to the landholders, and that payment often dwarfs other revenue generated by the land (for example, ranching), which creates an incentive for better stewardship. "There's intrinsic value in those resources," owner Rob Pacheco notes. "We're the mechanism that's converting that value into dollars."

One important mechanism for the transfer of value in ecotourism for HFT is to have exclusivity of access—not only to differentiate their brand, but to keep the landholder partner invested so that it doesn't turn into a tragedy of the commons. However, HFT believes what makes their program unique, what makes their clients willing to pay more for their services than for other competing operators, is the interpretative experience—that HFT offers them the best opportunity to connect to the resource. HFT prides themselves on transforming clients from "trophy-seekers" into stakeholders, who continue to return and provide financial and political support for protection of the resource.

The ideal corollary to exclusivity is for the revenue generated to go directly back to the resource. If exclusivity is based on the assumption that only certified sustainable operators are able to take tourists in because of the sensitivity of the resource, the most effective way to ensure the maintenance of the resource is to connect the financing back to it. In the case of Hawaii, this works well for operations on private lands, but for their visits within the public parks system HFT pays fees for permits to the state of Hawaii, and there is no guarantee that this revenue will go back to the specific park providing the benefits.

Monitoring and Verification

Little information exists in either the scientific literature or the certification material itself about what constitutes conformance to a standard (how many of the criteria need to be met) or how regularly a company will be audited to maintain its status. GTSC requires the application of "standard, transparent, and impartial verification procedures, and auditors who are technically competent in sustainable tourism and conformity assessment," and it follows ISO trends in verification and conformity assessment procedures. There are no published compliance statistics for Hawaii Eco-Tourism, but the most successful operators like HFT tend to go above and beyond their standards.

Effectiveness

There is a great need to evaluate the performance of ecotourism certification with respect to conservation, but the majority of publications on ecotourism have been related to impacts on social issues such as indigenous culture or distribution of tourism revenue. There have been very few quantitative studies measuring the impact of ecotourism on biodiversity, and even if the local effects of ecotourism on biodiversity are positive, there is still the consideration of the global and cumulative effects of increasing long-haul travel (if ecotourism is more expensive and thus caters more to wealthy and remote Westerners) on carbon emissions and thus climate change.

Ecotourism has been found to deliver significant social, environmental, and economic benefits in the Osa Peninsula, Costa Rica: reducing disparities in access to resources, providing financing for national parks and tree-planting programs, and resulting in higher and more stable earnings. CST-certified Lapa Rios Eco-Lodge showed an increase in forest cover on its property and nearby (within 5 km), reversing a trend of deforestation, while areas farther away continue to lose forest.

To return to our example of HFT, the dividends to nature and local communities have been many faceted. They are committed to maintaining several ongoing stewardship projects such as the restoration of unique Hawaiian wetlands in Kohala. In addition to the percentage of gross-revenues paid to landholders to incentivize further conservation practice, HFT contributes a portion of their net profits to the community through in-kind and outreach programs focused on conservation, including the Grow Hawaii Festival, Hakalau Forest NWR Open House, and Waikoloa Dryland Forest Initiative. Finally, beyond their certification with Hawaii Eco-Tourism, HFT has formed strategic partnerships with several local, national, and international organizations devoted to conservation and sustainable tourism.

Key Lessons Learned

Hawaii Forest and Trail provides an excellent example, and there are many others worldwide, of ecotourism that can enhance and conserve natural capital. Ecotourism at its best provides income for people stewarding the natural resource that supports the activity providing that income, and it builds a strong ethic in the tourists themselves to protect the place they have come to see. The role of certification in ecotourism is still too early to assess; it remains to be seen whether third-party certification will influence traveler preferences sufficiently to provide a competitive advantage over uncertified properties. However, the relationships built between operators like HFT and their clients may provide financial support over the long term that exceed any price differential on a single transaction. The importance of a third-party certifier may be greater in the next case study about supply chains, where consumers do not have as direct a relationship with the place or the producer.

Case 2. Supply-Chain Certification: Food and Fiber

The Problem

Growth in consumption of agricultural, forest, and ocean products strains the ecosystems within and surrounding where these products are grown or harvested. The size of large consumer-goods companies is on par with that of many nations—and therefore adjusting their practices or the practices of producers comprising their supply chains could make a huge impact. Consumers can pay firms to internalize the externalities of their production or harvest, by purchasing more product or paying higher prices for products with an eco-certification label.

The Ecosystem Service

There are many ecosystem services supporting and impacted by production landscapes and seascapes. Agriculture and forestry certification standards tend to focus on some combination of five targets: soil, water, agrochemicals, land cover, and biodiversity. Land cover can of course impact the other targets, determining whether vegetation intercepts pollutants before they enter the water and whether biodiversity can be maintained in a landscape, and it also provides cobenefits that are often assumed by the program, like carbon storage and sequestration for meeting climate mitigation goals.

Fisheries certification standards typically focus on four targets: stock status, impacts on nontarget species, effects on endangered and threatened species, and ecosystem integrity. The latter likely affects the first three, as coastal habitats such as mangroves and coral reefs provide important nurseries to maintain fishery populations; they also provide other important ecosystem services that are not necessarily specifically defined by the program, like tourism or storm surge protection.

Ecosystem Service Beneficiaries

As for the case of ecotourism certification, commodity certification supports two different levels of ecosystem service beneficiaries. Consumers benefit from the knowledge that they are supporting more sustainable systems (especially for more global benefits like carbon sequestration, or existence value of forest, or biodiversity conservation). People who live in the watersheds, landscapes, or coastal systems where the production or harvest has been improved benefit from more locally produced services like erosion control or water purification. Certified and noncertified farmers alike also benefit from services accrued within their production landscapes, such as enhanced pollinator availability or pest control.

Ecosystem Service Suppliers or Mediators

The supply of ecosystem services in this case is maintained or reduced by the producers in these production landscapes and seascapes—farmers, foresters, fishers. Rather than supplying ecosystem services, these actors may undermine them by contributing to problems of pollution or overharvest if they are poor managers.

Terms of the Exchange: Quid Pro Quo

Producers or consumer-goods brands pay for certification in exchange for using a seal of approval on their product, with the expectation that this leads to economic benefits (price premium or brand value) and ecological benefits (more sustainable production, for enhanced biodiversity and ecosystem services). The market for certification is rapidly growing but still occupies a relatively small share of overall production, with the exception of coffee (for which certification now accounts for 38 percent of all production). Other major sectors of certification adoption include cocoa (22 percent); palm oil (18 percent); tea (12 percent); wild-capture fisheries (10 percent); and growing production segments for aquaculture (4 percent); cotton (3 percent); sugar (3 percent); soybeans (2 percent); and bananas (3 percent) (Potts et al. 2014).

Mechanism for Transfer of Value

There are many models for certification, but the common mechanism is for producers to meet a series of standards, in exchange for the use of the certification seal in their marketing and promotion. As coffee is the most established commodity for certification, it provides an excellent case study for what is possible to achieve through this mechanism. One of the most successful certification programs for coffee is managed by Rainforest Alliance (RFA). More than 300,000 metric tons of RFA-certified coffee are produced annually around the world, representing 2 percent of the global market. The RFA certification requires performance standards such as the implementation of a management system for the farm, ecosystem protection (e.g., forest remnants and riparian vegetation), wildlife protection, water conservation, integrated crop management, soil conservation, and integrated waste management, along with several social dimensions.

Monitoring and Verification

A third party certifies that regulations are being followed, often requiring "improvement" to be demonstrated at each audit. These audits are almost uniformly conducted at the farm scale, despite many programs aspiring to have impact at the landscape scale (such as biodiversity protection).

Effectiveness

There are many motivations for certification. Companies may hope it will justify a higher price from consumers supporting products more aligned with their values, enhance brand reputation, satisfy corporate sustainability commitments, reduce operational risk, or provide greater access to market. In comparison to ecotourism, the ecological benefits of commodity certification have been well studied. In fact, a 2015 review of thirty scientific studies investigating the ecological effectiveness of certification showed that a considerable majority (twenty-three) of them found a positive effect for services or biodiversity (Chaplin-Kramer et al. 2015). Coffee and seafood show the most positive results, while the effects of timber are mixed. The most commonly measured outcome is forest cover, followed by biodiversity, but no studies have documented impacts on soil and water.

Continuing to use RFA-certified coffee as an example, certification has had especially beneficial results in Latin America and Africa. RFA-certification has led to enhanced tree cover and greater landscape connectivity in the Colombian eastern Andes (Rueda et al 2015). Similar effects have been seen in Ethiopia, where RFA-certification increased the probability of forest conservation by 20 percent relative to forest coffee areas lacking certification (Takahashi and Todo 2013). This effect was even stronger (nearly 30 percent) with lower-income producers, suggesting that certification significantly impacts the behaviors of economically poor producers, motivating them to conserve the forest.

The other aspect of effectiveness of the program is economic benefit. While it is not evident that producers consistently receive a price premium for selling under certification, they benefit in many other ways. In Peru, RFA farmers did receive a small price premium, but its economic importance was far overshadowed by the increase in yields that came through technical assistance in this program (Barham and Weber 2012). In Colombia, the initial 40 percent premium paid for RFA-certified coffee was substantially reduced over time and yet farmers remained in the program, showing that while the promise of a premium for their coffee is often the reason why many certified farmers decide to join a certification program, it is rarely the reason they remain (Rueda and Lambin 2013). Incentives to remain in certification cooperatives include increased access to capital, technology, or knowledge, along with social benefits such as developing stronger family and community ties, improving treatment to workers and their overall quality of life, and playing a leadership role in their community.

Key Lessons Learned

Commodity certification is growing in popularity for both consumers and business, and has been linked to noteworthy improvements in biodiversity and ecosystem services for at least certain commodities (such as coffee). It also provides an important mechanism for consumer goods firms to meet their sustainability commitments without having full custody over their whole supply chain. Unlike for ecotourism, where the consumers interact directly with the producers, consumers of commodities are often many steps removed from the producers. Certification provides an important mechanism for incentivizing better production despite that lack of traceability.

It is less well established that producers adopting certification standards benefit financially, and even when price premiums have been instituted, they tend to diminish over time or with market fluctuations. However, it appears that many other social benefits accrue to producers from participating in certification programs, which suggests that these schemes are durable mechanisms for securing biodiversity and ecosystem services in production landscapes and seascapes.

Impact Investing for Ecosystem Services

Impact investing is a second type of market-based mechanism that can finance conservation of ecosystem services. In its broadest form, impact investing aims to generate financial returns in combination with social and/or environmental benefits—the so-called double- or triple-bottom line. When applied to ecosystem services, investors invest in a financial vehicle that spends funds in order to generate both cash flows and secure or enhance ecosystem services (fig. 10.2).

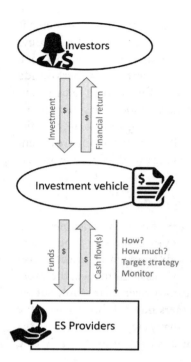

Figure 10.2. With impact investing, investors make financial contributions to an investment vehicle (e.g., bond, investment fund), which then provides funding to ecosystem service providers. The ecosystem service providers generate cash flows back to the investment vehicle, which provides financial returns back to the investors. The investment vehicle determines where and how to invest in securing or enhancing ecosystem services, and monitors project compliance.

The financial returns go back to the investors, while nonfinancial returns benefit the environment and society more generally. The impact investing model can be applied to a diversity of financial vehicles, from bonds to private equity funds. Some investors and vehicles aim to achieve risk-adjusted market rates of return, whereas others may accept a lower (also called "concessionary") rate of return in exchange for the environmental and social cobenefits.

A 2016 report by Credit Suisse projects that the conservation finance market, which includes impact investing for ecosystem services and other environmental benefits, could reach US$200–400 billion in private finance from institutional investors and high-net-worth individuals by 2020 (Huwyler, Käppeli and Tobin 2016). Interest in conservation finance has been bolstered by low interest rates and the lack of more attractive investment options. Conservation finance is also potentially attractive to investors as a means of diversification, because its returns are not expected to be correlated with other asset classes.

Many of the ecosystem service mechanisms covered in other chapters can also create opportunities for private finance. For example, regulatory-driven mitigation (chapter 7) creates opportunities for investment in mitigation banks, where large investments are needed up front to establish banks, and profits are generated as credits are sold. Eco-certification (case 2, this chapter) of commodities such as timber can also provide opportunities for private finance, with initial funding needed to change operation practices and attain certification, and profits generated over the longer term once production is underway.

Here, we focus on two case studies: the Water Sharing Investment Partnership (WSIP) in the Murray-Darling Basin in Australia as an example of private equity, and the District of Columbia Water and Sewer Authority's Environmental Impact Bond as an example of debt-based impact investing. Private investment that blends financial and conservation goals has existed in the United States for nearly twenty years, but investing specifically for ecosystem services is a newer and narrower subset of impact investing. A number of other promising examples are early in their implementation (see for example box 10.1). With a few more years of progress, it should be possible to learn from and evaluate the success of a wider array of investments.

Case 3. The Murray-Darling Basin Water Sharing Investment Partnership

The Murray-Darling Basin is known as Australia's food bowl (figure 10.3). The region contains 20 percent of the country's agricultural land area and produces one-third of its food supply, including nearly AUS$7 billion in irrigated produce. With an arid climate, rainfall varies greatly year to year and water losses to evaporation are high, which poses a challenge to meeting agricultural and other demands for water. Australia experienced a record-breaking drought from 1997 to 2009. Although the drought took a toll on agricultural production, the Basin's water market, which allows farmers to trade water rights, mitigated the economic impact: Water went to high-value crops and perennial crops that could not be fallowed, while low-value annual crops were let fallow until water availability increased. However, the drought also took a toll on wetlands, and freshwater and estuarine ecosystems, whose allocations of water were reduced or eliminated during this period of water scarcity. The challenge remained of how to efficiently allocate water to support agriculture in the region, while also maintaining water-dependent ecosystems and the

Box 10.1

The Forest Resilience Bond: Connecting Private Capital to Restoration Projects That Reduce Fire Risk and Provide Cobenefits

Benjamin P. Bryant, Zachary Knight, Phil Saksa, and Nick Wobbrock

In the western United States, the incidence of large, high-intensity fires has steadily increased over the past fifty years. A century of fire suppression has created unnaturally dense forest conditions that are more fire-prone than forests of the past. This density, when combined with an increasingly hotter and drier climate, has created the tinderbox conditions that made 2017 the most expensive fire season in US history, costing US$2.7 billion or 56 percent of the annual budget of the United States Forest Service (USFS). Unfortunately, this is a new normal, as these fires have incurred billions of dollars annually in fire suppression, postfire remediation, and infrastructure replacement costs, in addition to lives lost, damaged watersheds, and enormous carbon emissions.

Forest restoration can simultaneously improve the resilience of landscapes to fire, as well as make it easier to manage fire where necessary to protect people and property. In this context, restoration includes fuels reduction through thinning of small-diameter trees and undergrowth, prescribed fire treatments, and other activities that make fire less ecologically damaging and easier to control within a watershed. Forest restoration confers a variety of benefits that accrue to a diverse set of stakeholders beyond landholders—some directly from the restoration action itself, and some from the reduction in fire risk. These stakeholders can include (1) water and hydropower utilities benefiting from a more secure water supply; (2) utilities, companies, and communities benefiting from reduced fire risk to housing and infrastructure; (3) federal, state, and local governments benefiting from reduced firefighting and postfire rehabilitation costs; (4) populations benefiting from reduced exposure to unhealthy smoke; and (5) local residents and tourists benefiting from recreation opportunities.

The USFS has identified 58 million acres at very high risk of severe fire and in need of ecological forest restoration, but the agency lacks the funds to implement at the pace and scale necessary, despite the significant benefits.

The Forest Resilience Bond (FRB) provides a novel strategy to address this gap. Developed by Blue Forest Conservation and the World Resources Institute, it connects private capital to certified implementation partners, to increase the pace and scale of restoration beyond what government funding can achieve. Funds are provided up front by private capital, and are then returned with interest via annual contracted cash flows from the land manager and other beneficiaries.

The process for a successful FRB is captured at a high level in figure 10.A.1, and includes the following steps: First, a project must be planned and permitted, typically by the land manager, including diverse stakeholder input. Second, an economic assessment that estimates the benefits to downstream and watershed stakeholders is conducted. Third, stakeholders confirm their interest in the project and commit to contracted payments based on the benefits they receive. Fourth, potential investors who will be expected to provide up-front capital in exchange for being repaid over time are identified. Each stakeholder contributes a portion of the project costs, and the combination of all stakeholder contracts covers the cost of the project and a modest return for investors. These payments can be made on a cost-share basis or based on agreed-upon environmental outcomes that are measured following restoration. Fifth, investment dollars are used to fund a nonprofit implementation partner that has a strong connection with the forest, the stakeholders, and the community. This partner hires local contractors and ensures work is completed in a timely manner and according to prescriptions. As work is completed, verification includes USFS oversight and third-party monitoring from academic project partners.

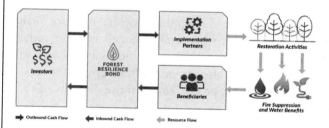

Figure 10.A.1. The Forest Resilience Bond structure.

As of November 2018, the first pilot FRB has been signed, with restoration activities underway on national forest system land in the North Yuba watershed in California. Besides the USFS, the project brings together multiple funders, with the National Forest Foundation as the main implementer and the Yuba Water Agency as the main contracted beneficiary. The agency has signed cost-share contracts to repay a portion of the restoration cost over time from utility revenues, supplementing USFS payments, and helping to provide a return for investors. Monitoring during implementation includes satellite-based assessment of changes in the forest water balance following restoration. As implementation and measurement of benefits improve over time, so does the ability to characterize and shift risk and return according to preferences of different participants, through performance-based contracts. Together, these improvements can encourage additional beneficiaries and investors to participate in FRB projects.

Collective action often stalls when benefits are enjoyed by a diffuse set of stakeholders. By developing flexible and innovative contracts, where each beneficiary can contract along their required terms, timelines, and bureaucratic constraints, the Forest Resilience Bond provides an innovative solution to meet the enormous need for forest restoration that land managers have identified in the American West.

services they provide. To address this challenge, The Nature Conservancy (an international nongovernmental organization), the Murray Darling Wetlands Working Group Ltd. (an Australian not-for-profit company), and Kilter Rural (an Australian asset management firm) collaborated to create a Water Sharing Investment Partnership (WSIP) called the Murray-Darling Basin Balanced Water Fund. By buying water rights, then leasing some of the water back to farmers and donating the remaining water for environmental purposes, the fund aims to secure water for agriculture, restore threatened wetlands, and generate returns for investors.

The Ecosystem Service

The allocation of water influences the provision of ecosystem services from agriculture and ecosystem services from wetlands. Water that is allocated to agriculture promotes the production of food and fiber. The Murray-Darling Basin produces more than 50 percent of Australia's irrigated produce, including cotton, hay, livestock, almonds, and stone fruit. Water allocated to wetlands promotes the conservation of biodiversity, with a specific focus on birds, wildlife, and threatened species. Wetlands also provide cultural ecosystem services for Australia's Aboriginal (indigenous) people, supporting a diversity of social, cultural, and spiritual values.

Ecosystem Service Beneficiaries

Farmers and consumers benefit from the water that is allocated to agriculture. The water that is allocated to wetlands provides benefits globally to those who care about biodiversity conservation, as well as locally to Aboriginal communities. Water dedicated to the environment also provides greater system sustainability, which ultimately should increase the sustainability and longevity of agricultural production in the region.

Ecosystem Service Suppliers

By holding water rights and deciding how they are allocated between agriculture and environmental uses, the WSIP functions as the ecosystem service supplier. The farmers and wetland landowners who receive water also contribute to the provision of ecosystem services.

Terms of the Exchange: Quid Pro Quo

The Murray-Darling Basin Water Fund buys water rights and each year leases between 60 and 90 percent of its water to farmers for agricultural purposes.

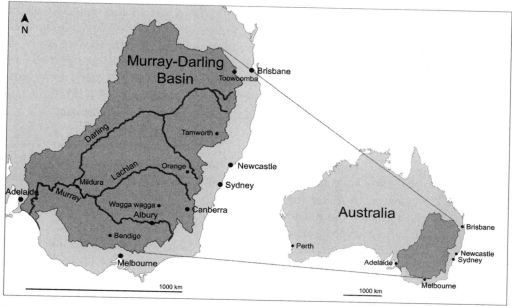

Figure 10.3. The Murray-Darling Basin produces one-third of Australia's food supply. Located in an arid region susceptible to drought, allocating water between agricultural uses and the environment is a challenge. The Murray-Darling Basin Balanced Water Fund aims to secure water for agriculture and increase the allocation of water to environmental purposes while generating a profit for fund investors through its acquisition and management of water rights.

Profits from leasing water rights go back to fund investors. The fund donates the remaining 10–40 percent of its water holdings to wetland restoration, with a focus on wetlands or floodplain forests with high conservation value on private land that would not otherwise receive water from government programs.

The fund raised about US$20 million from investors in its first round of funding, which closed in December 2015, and aims to scale up to US$75 million in four years (Kitney 2016). As of October 2017, the fund owns the rights to 8.5 gigaliters of water and is aiming for 7–9 percent returns over the long term (Kilter Rural and TNC 2017).

The fund has contributed to several environmental watering events, including: one in April 2016, when 950 megaliters of water went to wetlands on

Tar-ru Traditional Lands in New South Wales, and another in the Goulburn River floodplain in September 2017 (Kilter Rural and TNC 2017). In the 2016 watering event, the WSIP provided capacity for the Murray Darling Wetlands Working Group to plan, administer, and deliver water that was provided by the government (Commonwealth Environmental Water Holder), rather than providing the water itself. The water went to wetlands on what is currently state land that will be returned to Aboriginal ownership and management.

Mechanism for Transfer of Value

The Water Sharing Investment Partnership (WSIP) has two main parts: a Balanced Water Fund and Environmental Water Trust. The Balanced Water Fund is managed by Kilter Rural. Fund investors include foundations, mission-driven pension funds, and high-net-worth individuals; the National Australia Bank also provided a loan to the fund.

The fund buys permanent water entitlements, then leases a portion (60–90 percent) of these entitlements back to farmers in dry years. It donates 10–40 percent of water entitlements to the Environmental Water Trust, which is a tax-deductible, not-for-profit organization. More water is allocated for environmental purposes in wet years when demand from agriculture is lower, which also helps to mimic natural patterns of water provision to wetlands.

Finally, a Scientific and Cultural Advisory Committee helps the Environmental Water Trust achieve its environmental and cultural objectives. These objectives include conservation benefits such as improved health of vegetation and threatened species, as well as social and cultural benefits to Aboriginal communities through restoration of culturally important sites.

This WSIP is enabled by the existing water rights market in the Murray-Darling Basin, as well as by the conditions associated with water rights. In Australia, water is delivered to the holders of water rights proportionally based on water availability. So, for example, in a dry year, water rights holders might all receive only 75 percent of their water entitlements. This is in contrast to the western United States, where those with the oldest rights receive their full allocation of water first. Since 2007, it is possible to own and trade water rights without owning land in the Murray-Darling Basin.

A monitoring program to evaluate environmental and cultural outcomes has been set up, and results are reported annually. The fund is also supporting training for people from indigenous communities to enable their participation in monitoring.

Effectiveness

On the environmental side, as of June 2018 the WSIP has provided financial support for water events starting in 2016, as well as donations of water allocations in 2017 and 2018. As of January 2019, the fund had an annualized return of 14.93 percent since inception (Kilter Rural 2019), exceeding its target of 7–9 percent returns. Given that the fund has been in operation only since 2015, its long-term effectiveness across its financial, environmental, and social/cultural objectives remains to be seen.

Key Lessons Learned

The example of the Murray-Darling Basin Water Sharing Investment Partnership highlights the roles for, and importance of, multiple types of actors and institutions in impact investing. In this case, the government plays a key role in enabling the WSIP mechanism by setting water rights and water market conditions, for example, by allowing entities to own and trade water rights without owning land in the Basin. Private investors, including some with philanthropic aims, provide the water fund with the capital needed to purchase water entitlements, while not-for-profit companies/nongovernmental organizations like the Environmental Water Trust are critical to carrying out the WSIP's environmental and social objectives on the ground.

Based on its experience in the Murray-Darling Basin and with water rights and water markets globally, The Nature Conservancy has proposed a number of enabling conditions that are needed for successful implementation of the WSIP model. These include the existence of legally defined and enforceable water rights, the ability to trade conserved water, the ability to transfer permanent water rights and short-term water access, buy-in from the agricultural community and other stakeholders, and monitoring and enforcement to ensure compliance (Richter 2016). These can be used to identify where WSIPs could provide an opportunity for private investment to help balance the allocation of water across multiple societal benefits, including agricultural production and biodiversity conservation.

Case 4. DC Water Environmental Impact Bond

The Problem

The area of paved and impervious surface in Washington, DC—and many other cities—has grown, preventing rainfall from being absorbed into the soil (see chapter 7, case 1, for more detail). As a result, the city's combined sewage/stormwater runoff system overflows regularly, sending 10 million m³ of runoff and untreated sewage flowing into local water bodies, and ultimately the Chesapeake Bay every year. This has degraded the quality of these water bodies. The overflows are a violation of the federal Clean Water Act. In 2005, the US Environmental Protection Agency, Department of Justice, District of Columbia, and District of Columbia Water and Sewer Authority (DC Water) agreed to allow DC Water to address the problem by constructing a US$2.6 billion tunnel system that would capture overflows and allow them to be treated before entering waterways.

Since then, nature-based solutions to reducing stormwater flows have emerged as a viable option, and DC Water was able to modify its agreement with the federal government to allow the use of green infrastructure to reduce its combined sewer overflows. In 2016, DC Water issued an Environmental Impact Bond to fund construction of green infrastructure to reduce stormwater runoff. The bond includes an innovative pay-for-performance structure in which payments to investors depend on the level of success achieved in reducing stormwater runoff.

The Ecosystem Service

Under normal circumstances, stormwater and sewage flow together in DC's combined sewer system, and this wastewater is treated before being discharged into local water bodies. In heavy rainfall events, the amount of stormwater and sewage generated exceeds the sewer system's capacity, and excess wastewater is discharged without treatment. Green infrastructure—such as rain gardens and green roofs—contributes to reducing these overflow events by increasing infiltration and slowing storm flows. This reduces the likelihood of the sewer system becoming inundated after a storm, with the resulting release of untreated sewage. If green infrastructure is successful at reducing stormwater, it is possible that the size (and therefore cost) of the planned additional tunnel system could be reduced. Green infrastructure also provides cobenefits that the tunnel system would not, including aesthetic values, recreational opportunities, and health benefits associated with green spaces.

Ecosystem Service Beneficiaries

The improved water quality from avoided combined sewer overflows benefits everyone downstream using the DC rivers and the Chesapeake, particularly those involved in recreational and commercial fishing and water-based tourism in the area. DC residents pay for the stormwater regulation service indirectly, through the portion of their taxes that go toward repaying bond investors. If green infrastructure proves to be more cost effective than the planned tunnel system, DC residents could additionally benefit from reduced costs for financing that.

Ecosystem Service Suppliers
The ecosystem service supplier is DC Water, which is constructing the green infrastructure that provides stormwater regulation services.

Terms of the Exchange: Quid Pro Quo
DC Water is using the proceeds from the sale of the bond to construct green infrastructure (specifically, permeable pavement and rain gardens) on 8 hectares of land, in order to pilot-test its effectiveness (Valderrama 2016). DC Water pays bondholders annual interest and will repay the principal after thirty years. In 2021, after five years, there may also be performance-based payments between DC Water and bondholders (see *Mechanism for Transfer of Value* for more details). Stormwater runoff will be monitored at the site before and after construction of green infrastructure. The amount of payment depends on how successful the green infrastructure proves to be at reducing stormwater flows.

Mechanism for Transfer of Value
DC Water's funding for constructing green infrastructure comes from the proceeds of the US$25 million bond, which was sold to Goldman Sachs and Calvert Foundation, a nonprofit investment firm, now Calvert Impact Capital, (Quantified Ventures 2016). The bond is a thirty-year bond with a 3.43 percent coupon rate (annual interest). This rate is comparable to the historic rate for other thirty-year bonds that DC Water has issued.

Under the bond's pay-for-performance structure, however, payment back to investors depends on the environmental success of the green infrastructure at the five-year mark (Martin 2017). If the green infrastructure meets expectations, then no performance-based payment occurs, and the bond functions as a conventional bond. DC Water and its partners modeled a range of possible outcomes and used the middle 95 percent of those outcomes as the expected range. If the green infrastructure outperforms expectations, DC Water will pay an additional US$3.3 million to bond investors, increasing their return to 6.4

percent. In this case, the green infrastructure would prove to be more effective at reducing runoff than expected, and so DC Water would be able to save money when it scales by reducing the amount of green infrastructure needed, and possibly grey infrastructure as well, to manage the same amount of stormwater. On the other hand, if green infrastructure drastically underperforms, the investors will pay DC Water US$3.3 million, essentially covering the cost of DC Water's interest payments for the first five years. This serves as a sort of insurance policy for DC Water, should its green infrastructure pilot fail.

Ultimately, this pay-for-performance structure distributes the risk associated with green infrastructure between DC Water and bond investors. If the pilot succeeds, DC Water expects to spend US$90 million to construct 120 hectares of green infrastructure for stormwater management.

Effectiveness

The effectiveness of the permeable pavement and rain gardens at reducing stormwater flows will be independently verified. All water from the pilot site is channeled into a gauged pipe to allow for direct measurement. Stormwater flows have been measured for twelve months before installation of green infrastructure as a baseline and will be compared against twelve months of data after the green infrastructure is built. The 95 percent range for the expected outcome is meant to cover, among other things, potential differences in weather between years that could influence the results. Results should be available by 2021.

Key Lessons Learned

As with the Murray-Darling Basin Water Sharing Investment Partnership, this example of impact investing shows the complementary roles that government and private interests can play. Washington, DC's interest in using green infrastructure for stormwater management stems from the regulations set by the federal government under the Clean Water Act. Private investors then provided the up-front capital DC Water needed to pilot-test the use of green infrastructure in exchange for anticipated returns over the next thirty years.

The performance risk associated with green or nature-based infrastructure has been one major limitation to its adoption over gray or engineered infrastructure. The pay-for-performance model employed by the DC Water Environmental Impact Bond provides a way of transferring much of that risk from governments, with limited funds and a need to deliver results cost effectively, to investors who are willing to take on risk when it is associated with adequate financial returns. Such a model requires the ability to sufficiently quantify project risk up front and to measure project performance.

Conclusions

One of the main advantages of ecosystem service approaches is in making the previously hidden values of nature more explicit and transparent, so they are not assumed to be valueless. The market-based mechanisms highlighted here provide a way of correcting market failures: where the traditional markets are not adequately capturing the value provided by ecosystems, new markets are set up to allow consumers to bear some of the costs or risks of securing or enhancing ecosystem services. Ideally, these mechanisms will not only expand the funding possibilities for conservation beyond government and philanthropic sources, but also create a more sustainable system for financing natural capital security, independent of political swings or the whims of donors. The two types of market-based mechanisms highlighted here, eco-certification and impact investing for ecosystem services, represent a range of options that are currently commonplace enough that some models of success already exist.

As the idea of ecosystem services becomes more mainstream, other market-based mechanisms may arise. For example, Swiss Re, a reinsurance company, is working with The Nature Conservancy, the Mexican government, and other partners to develop an insurance policy for the coral reefs that protect the coastlines of Cancún, Mexico (Harvey 2017). The hope is that this will provide a model for insuring natural capital and ecosystem services elsewhere.

Market-based mechanisms will likely work particularly well when combined with government regulations requiring certain levels of environmental quality or human well-being supported by natural capital. While in many cases it is the role of government to ensure public goods are provided, market-based mechanisms can be an efficient way of delivering them. There is great growth potential for these approaches to provide sustainable solutions to financing natural capital, if we can properly attribute who is benefiting from nature, and to what degree those beneficiaries have the capacity and incentive to pay for the continued provision of those benefits.

Key References

Barham, Bradford L., and Jeremy G. Weber. 2012. "The economic sustainability of certified coffee: Recent evidence from Mexico and Peru." *World Development* 40, no. 6:1269–79.

Buckley, Ralf. 2009. "Evaluating the net effects of ecotourism on the environment: A framework, first assessment and future research." *Journal of Sustainable Tourism* 17, no. 6:643–72.

Chaplin-Kramer, Rebecca, Malin Jonell, Anne Guerry, Eric F. Lambin, Alexis J. Morgan, Derric Pennington, Nathan Smith, Jane Atkins Franch, and Stephen Polasky. 2015. "Ecosystem service information to benefit sustainability standards for commodity supply chains." *Annals of the New York Academy of Sciences* 1355, no. 1:77–97.

Harvey, Fiona. 2017. "Mexico launches pioneering scheme to insure its coral reef." *The Guardian*, July 20, 2017. https://www.theguardian.com/environment/2017/jul/20/mexico -launches-pioneering-scheme-to-insure-its-coral-reef.

Huwyler, Fabian, Jürg Käppeli, and John Tobin. 2016. *Conservation Finance from Niche to Mainstream: The Building of an Institutional Asset Class*. Credit Suisse and McKinsey Center for Business and Environment.

Kilter Rural. 2019. *The Murray-Darling Basin Water Fund Investor Update, January 2019*. Bendigo, Australia: Kilter Rural.

Kilter Rural and TNC. 2017. *The Murray-Darling Basin Balanced Water Fund Information Memorandum*. Kilter Rural. http://kilterrural.com/news-resources/murray-darling-basin -balanced-water-fund-information-memorandum.

Kitney, Damon. 2016. "Heavyweights invest $100m in Murray-Darling water." *The Australian*, April 4, 2016.

Martin, Abby. 2017. "A pioneering environmental impact bond for DC water." *Conservation Finance Network*, January 2, 2017. http://www.conservationfinancenetwork.org/2017/01/02 /pioneering-environmental-impact-bond-for-dc-water.

Medina, Laurie Kroshus. 2005. "Ecotourism and certification: Confronting the principles and pragmatics of socially responsible tourism." *Journal of Sustainable Tourism* 13, no. 3:281–95.

Potts, Jason, Matthew Lynch, Ann Wilkings, Gabriel A. Huppé, Maxine Cunningham, and Vivek Anand Voora. 2014. *The State of Sustainability Initiatives Review 2014: Standards and the Green Economy*. Winnipeg, MB: International Institute for Sustainable Development.

Quantified Ventures. 2016. *DC Water's Green Infrastructure Environmental Impact Bond Overview*. http://www.quantifiedventures.com/s/DC-Water-EIB-Overview.pdf.

Richter, Brian. 2016. *Water Share: Using Water Markets and Impact Investment to Drive Sustainability*. Washington, DC: The Nature Conservancy.

Rivera, Jorge. 2002. "Assessing a voluntary environmental initiative in the developing world: The Costa Rican Certification for Sustainable Tourism." *Policy Sciences* 35, no. 4:333–60.

Rueda, Ximena, and Eric F. Lambin. 2013. "Responding to globalization: Impacts of certification on Colombian small-scale coffee growers." *Ecology and Society* 18, no. 3.

Rueda, Ximena, Nancy E. Thomas, and Eric F. Lambin. 2015. "Eco-certification and coffee cultivation enhance tree cover and forest connectivity in the Colombian coffee landscapes." *Regional Environmental Change* 15, no. 1:25–33.

Takahashi, Ryo, and Yasuyuki Todo. 2013. "The impact of a shade coffee certification program on forest conservation: A case study from a wild coffee forest in Ethiopia." *Journal of Environmental Management* 130:48–54.

Valderrama, Alisa. 2016. "Pay for performance meets green infrastructure." NDRC. https:// www.nrdc.org/experts/alisa-valderrama/pay-performance-meets-green-infrastructure.

CHAPTER 11

Multilateral and Bilateral Mechanisms

Rick Thomas

Reducing Emissions from Deforestation and Land Degradation (REDD+) has emerged as one of the most significant and best-funded policy initiatives in recent years to address climate change through land use, particularly in the tropics. By enabling developed countries to invest in proper forest management techniques, REDD+ creates a mechanism whereby forests can be more valuable as carbon sinks than chopped down. This chapter examines the operation of a multilateral-funded REDD+ project in Indonesia and a bilateral-funded REDD+ project in Brazil.

Policymakers have long understood the critically important role forests play in providing the ecosystem service of carbon sequestration. While many hoped the Clean Development Mechanism of the Kyoto Protocol would meaningfully address deforestation, credits were granted only for afforestation and reforestation. No credit was provided for reducing rates of deforestation. This policy gap provided the impetus for development of the Reducing Emissions from Deforestation and Forest Degradation (REDD) policy. Through the REDD framework, developed countries provide funds to developing countries for reductions in greenhouse emissions from deforestation and forest degradation, as well as for the enhancement of forest carbon stocks and forest cover. Billions of dollars have been spent promoting this policy approach. In 2010, at the Conference of the Parties to the United Nations Framework on Climate Change (COP 16 to UNFCC), REDD became known as REDD+, reflecting the additional goals of poverty alleviation, biodiversity conservation, and sustaining vital ecosystem services.

This chapter focuses on payments for climate regulation by the UNFCCC's REDD+ program through two case studies. The first case in Indonesia provides an example of multilateral funding in which funding has been pooled from multiple governments. A similar approach may be seen in chapter 13, describing the Forest Carbon Partnership Facility Carbon Fund in Costa Rica. The second case focuses on the Amazon Fund, a bilateral funding mechanism between Brazil and Norway.

Case 1. Multilateral Agreements: Forest Carbon Partnership Facility Readiness Fund in Indonesia

The Problem

Indonesia is home to the third largest tropical forest (after that in the Brazilian Amazon and in the Democratic Republic of Congo), and its peatlands have the largest carbon stocks in the tropics. Anthropogenic fires within these forests and peatlands, as well as the conversion of forests for uses such as oil palm plantations, have made the country a global leader in land-based emissions. Over a twelve-year period from 2000 to 2012, the country lost over 6 million hectares (ha) of its forest land, with deforestation rates increasing by 47,600 ha per year on average over that same time period. In total, this degradation and deforestation accounts for about 85 percent of the country's carbon emissions. Moreover, Indonesia contains high floral and faunal biodiversity, including 10 percent of the world's plants, 12 percent of the world's mammals, 16 percent of the world's reptiles and amphibians, and 17 percent of the world's bird species, many now threatened with extinction.

While this habitat destruction is a global concern because of the carbon emitted, it also causes disproportionate harm to Indonesian citizens. Many communities in Indonesia rely on forests for their livelihoods. It is estimated that more than 74 percent of the country's poor depend on local ecosystem services for their livelihoods, services that are compromised as the land becomes degraded or deforested (UNORCID 2015). Smoke from the slow-burning peatlands has also been linked to health impacts such as respiratory disease and decreased productivity that in turn adversely impact the economy: it is estimated that fires cost Indonesia US$30 billion in 2015 (Petrenko, Paltseva, and Searle 2016). A report found that illegal logging cost the country over US$2 billion in lost revenue in 2006 alone (Harwell 2009).

The Ecosystem Services

Forests provide numerous ecosystem services. From a global perspective, forests' ability to store huge amounts of carbon helps mitigate the dangerous effects of climate change. On a more local level, forests help prevent soil erosion and nutrient loss, and can increase water availability by slowing stormwater runoff, thereby filtering the water and recharging aquifers.

A REDD+ study (UNORCID 2015) in the province of Central Sulawesi found that one hectare of forest prevents soil erosion equivalent to 6,538 kg/ha a year, which translates to an avoided cost of approximately US$30 ha/year when considered alongside avoided soil nutrient loss due to surface runoff. The study found that the economic value of soil erosion prevention across five key provinces ranges from US$2 to 81 million per year. The economic value of carbon sequestration and storage ranges within these provinces range from US$17 to 97 million and US$1.2 to 19 billion per year, respectively. The economic value of water augmentation ranges from US$435 million to 2.4 billion per year.

Ecosystem Service Beneficiaries

Beneficiaries are those whose livelihoods or quality of life would be compromised with continued deforestation. Communities that rely on local water sources for drinking and cooking would benefit since these waters become badly polluted with sediment and nutrient runoff following extensive forest clearing for agriculture. Local forests provide valuable ecosystem goods for rattan and many medicinal and other plants central to livelihoods and well-being. Additionally, because carbon is a global pollutant, the global community is a beneficiary of these ecosystem services, as it stands to be worse off should climate change effects amplify. In particular, communities vulnerable to harmful climatic events will benefit the most, such as inhabitants of coastal communities and low-lying islands.

Ecosystem Service Suppliers

Under the negotiations leading to creation of the Readiness Fund, Indonesia's government is the ecosystem services provider and the entity compensated for its actions. While this is a national approach, under REDD+ subnational entities—such as provinces—can be considered suppliers of the ecosystem service, and therefore receive compensation.

Terms of the Exchange: Quid Pro Quo

The Readiness Fund is meant to preserve and protect forest ecosystem services by helping developing countries prepare for participation in future, large-scale REDD+ initiatives. The Fund does this by assisting parties in adopting national REDD+ strategies; developing reference emission levels (RELs); designing measurement, reporting, and verification (MRV) systems; and setting up REDD+ national management arrangements, including proper environmental and social safeguards. A reference level is the amount of greenhouse gas emissions from deforestation and forest degradation that a country's new emissions will be evaluated against and is typically the historical average of the area being monitored over a given time frame. If a country's emissions post-REDD+ are lower than the reference level, it has achieved emissions reductions. Safeguards are strategies that ensure emissions reductions do not compromise other environmental or social values, such as biodiversity or the well-being of indigenous communities.

Countries interested in receiving money from the Readiness Fund first submit a Readiness Plan Idea Note (R-PIN). The R-PIN is meant to serve as a template outlining the minimum requirements that a country must meet to begin its REDD+ implementation. For example, in Indonesia where illegal logging is a main driver of deforestation, the R-PIN recommends significantly improving law enforcement and compliance to implement meaningful REDD+ strategies. Once the R-PIN is approved, the next step is for countries to prepare their Readiness Plan, which is a framework that sets a clear approach, budget, and schedule to undertake REDD+ activities. Once this is approved, countries are eligible for Readiness Funds to finance the capacity building set forth in the plan.

Mechanism for Transfer of Value

A country's progress toward the Readiness Fund is overseen by the Forest Carbon Partnership Facility (FCPF). The FCPF is a global partnership of businesses, individuals, and governments created in 2007 following the Bali Accords. Although the Readiness Fund is multilateral, meaning money can come from multiple private and public sources, to date only public payments from national governments have been received. The entities that have contributed to the Readiness Fund include the European Commission, Australia, Canada, Denmark, Finland, France, Germany, Italy, Japan, the Netherlands, Norway, Spain, Switzerland, the United Kingdom, and the United States. With the World Bank as the trustee overseeing these donations, the Readiness Fund has received US$365 million in commitments since its inception.

As of June 2018, Indonesia has received over US$4 million in start-up finance from the Readiness Fund. Disbursement of funds has occurred in the form of grants starting in 2011.

Monitoring and Verification

Before distributing funds, the FCPF reviews and assesses each country's Readiness Plan. In particular, the FCPF evaluates whether or not each plan will help the country meaningfully implement REDD+ strategies in the future. The FCPF determines this by verifying that the country has designed a rigorous national REDD+ strategy, identified accurate and appropriate reference emission levels, and created a system to monitor and verify the impact of future programs.

Additionally, the FCPF verifies that each plan has a high degree of consultation with civil society and indigenous communities. This stipulation can be satisfied if the country has developed participatory mechanisms to ensure that indigenous communities are meaningfully consulted during the formulation and implementation of their country's Readiness Plan and REDD+ Strategy and benefit from the country's capacity building.

Effectiveness

The Readiness Fund has successfully helped Indonesia and thirty-six other countries design and implement rigorous infrastructure for REDD+ programs. As of June 2016, Indonesia's REDD+ Strategy has been finalized, meaning that the country has successfully established its National Forest Monitoring System (NFMS), and determined its reference emission level, mechanism for measurement, reporting, and verification, and Safeguards Information System for REDD+ (SIS REDD+).

The Readiness Fund process has helped Indonesia identify areas of weakness in terms of forest management that the country would likely not have had the capacity to address on its own. Specifically, according to feedback from the FCPF, the country needed to invest more in capacity building and awareness of REDD+ for local communities and develop a clearer system of incentives and performance-based mechanisms to ensure meaningful progress toward emissions reductions.

While this process has helped Indonesia address these deficiencies, because the country has yet to submit results from its REDD+ strategies, the Readiness Fund's effectiveness in reducing deforestation and forest degradation is still unknown.

Key Lessons Learned

The case study of REDD+ financing in Indonesia highlights the importance and necessity of start-up funds for capacity building. Despite the widespread damage being done to Indonesia's forests, it is unlikely that meaningful progress toward forest management strategies would have occurred without this type of start-up assistance. This is predominantly because of corruption, the economic value that private actors can capture from deforestation, and the diffuse nature of the primary ecosystem service benefit (while services like soil erosion prevention can be internalized by the country, carbon storage, currently, cannot). An investigation into Indonesia's forest policy gives credence to this idea: the country established a moratorium on new permits to clear primary forests in 2011 only *after* Norway agreed to fund up to US$1 billion in deforestation projects. It is interesting to note however, that out of this amount, Norway has distributed only about US$108 million.

Because the majority of money available for forest carbon is results based, having rigorous systems in place like those that the Readiness Fund helps develop is a necessary prerequisite to meaningfully obtaining and quantifying emissions reductions.

Case 2. Bilateral Agreements: Amazon Fund (Norway-Brazil)

The Problem

The Amazon rainforest in South America is the largest on the planet and is estimated to contain over 390 billion trees. Covering 550 million ha, the massive forest extends through eight countries, a fact that often makes designing consistent forest management difficult. Diverse geographic features and government policies among these countries have resulted in a very uneven distribution of deforestation in the Amazon—nearly 80 percent of all deforestation of the forest has occurred in Brazil. From 1988 to 2006, deforestation rates in the Brazilian Amazon averaged 18,100 km²/year, peaking in 2004. As of 2017, about 768,935 km² of Amazonian forests had been cleared since 1970, much of it illegally (Butler 2017). The main drivers of the clearcutting have been to grow soybeans and raise cattle, two important sources of income for the country.

The Ecosystem Services

The sheer size of the Amazon results in global-scale ecosystem services. A 2007 study found that the Amazon stores some 86 billion tons of carbon, more than a third of all carbon stored by tropical forests worldwide; it is estimated that in a single year, the forest is capable of absorbing 1.5 gigatons of carbon dioxide from the atmosphere. In addition to greenhouse gases, the forest extraction of soil water by tree roots recycles 25–50 percent of rainfall in the Amazon Basin, a phenomenon key to global atmospheric circulation.

Ecosystem Service Beneficiaries

Tens of millions of people live within the Amazon, including over 400 different indigenous communities, many of which rely on the services of the forest to sustain their way of life. This is additional to the global beneficiaries of reduced deforestation, as described in the *Ecosystem Service Beneficiaries* (case 1) section for the Indonesia example.

Ecosystem Service Suppliers

Under the Amazon Fund, the Brazilian government is the party responsible for providing the ecosystem services of carbon sequestration and storage.

Terms of the Exchange: Quid Pro Quo

In order for Brazil to receive payments through the Amazon Fund, the country has to develop a number of initiatives to ensure meaningful reduction in greenhouse gas emissions. These initiatives include a national strategy plan that addresses drivers of deforestation and forest degradation, land tenure issues, forest governance issues, gender considerations, and safeguards. Similar to requirements under the Readiness Fund, Brazil must also implement a national forest reference emission level, a robust and transparent national forest monitoring system for REDD+ activities, and a system for providing information on how REDD+ safeguards are being addressed. Finally, Brazil must also submit biennial reports on emissions and a Most Recent Summary that details how the safeguards are being addressed under REDD+.

Mechanism for Transfer of Value

Unlike the Readiness Fund discussed in the previous case study, the Amazon Fund is a bilateral payment mechanism, meaning that Brazil negotiates a contract with a single donor, in this case, Norway. (Other bilateral negotiations that are part of the Amazon Fund include arrangements with Germany and the Brazilian petroleum company, Petrobras.) Following negotiations, funds are transferred to the Brazilian Development Bank, the entity managing the Amazon Fund, and become available after the verification of predetermined results. As of 2018, Norway has donated over US$1.1 billion into the Amazon Fund. Although the funds are meant for forest protection measures, Norway and Brazil's contract does not require Brazil to verify how it is using payments from the Amazon Fund.

Monitoring and Verification

Similar to other results-based payment schemes, Brazil's emissions reductions are evaluated against a reference level based on a historical average. However, as negotiated between Brazil and Norway, every five years a new reference level is set at the last five years' average level, requiring Brazil to consistently improve its strategies to address deforestation over time. While the emission reductions themselves are independently verified, under the contract, Brazil's plans to address safeguards are not monitored or evaluated in any way.

Following the country's emission reductions verification, all pertinent documents and results are posted to the Lima Information Hub, an online database organized by the UNFCCC to aid in transparency.

Effectiveness

According to the Lima Hub, from 2006 to 2010, Brazil successfully reduced emissions totaling over 511 million tons of carbon equivalent per year and over 277 million tons per year from 2010 to 2015: altogether 6.125 gigatons of carbon equivalent. To date, Brazil is the only country to submit verified results to the Lima Hub. Once other countries have verified reductions, their results will be posted online as well.

Key Lessons Learned

Much of Brazil's emission reductions resulted from rigorous government programs, which were driven and made more robust as a result of international assistance. This highlights the importance of designing the international protocol to support parties' national policies to address deforestation rather than undermine them.

Brazil's success in reducing its emissions from deforestation has helped generate other key lessons that can guide future REDD+ programs. For example, Brazil found that its REDD+ initiatives were most successful when integrated into cross-sectoral policies such as agriculture, forestry, infrastructure, and environmental policies. By integrating REDD+ initiatives, as opposed to keeping them isolated, each sector is forced to consider their contribution to greenhouse gases from deforestation, and the potential role they can play in mitigation (Viana et al. 2012). Another key insight lies in the critical importance of providing assistance for capacity building to successfully implement REDD+ initiatives. Capacity building should not be limited to simply helping put infrastructure in place but include educational outreach to help inform local communities of the issues and how they can contribute, which Brazil addressed through workshops and communication campaigns (Viana et al. 2012).

Brazil's success emphasizes how the most effective REDD+ initiatives focus on community-oriented forest management. By including local and indigenous communities that know the forests in the design of management plans, they are more likely to adhere to the plans, ultimately making the strategies more pragmatic and robust. These findings have been used to help inform REDD+ projects in several African countries, including Cameroon, Gabon, the Central African Republic, Democratic Republic of Congo, and Republic of Congo.

Conclusions

The REDD+ case studies illustrate the policy mechanism of conditional international aid. Development institution or national funds provide assistance so that important forested developing countries can put in place the local and national institutions and regulatory frameworks to both reduce deforestation and verify that the forest changes are really happening on the ground. These so-called REDD+ Readiness efforts demonstrate that directed investment can lead to significant changes in forest management over a short period of time.

While REDD+ Readiness funding is generally considered a success to date, such approaches have an obvious vulnerability. There is always risk of losing these benefits should local or national forest management significantly change course. A key question is whether the reduced deforestation will continue if funding stops. Ensuring the long-term provision of ecosystem services under REDD+ will require careful attention to the creation of institutions and funding mechanisms that are sustainable if the foreign income flows shrink in the future or stop altogether.

Key References

Butler, Rhett. 2017. "Calculating deforestation figures for the Amazon." Last modified January 27, 2017. https://rainforests.mongabay.com/amazon/deforestation_calculations.html.

Harwell, Emily. 2009. *"Wild Money": The Human Rights Consequences of Illegal Logging and Corruption in Indonesia's Forestry Sector.* Human Rights Watch. https://www.hrw.org/re port/2009/12/01/wild-money/human-rights-consequences-illegal-logging-and-corrup tion-indonesias.

Margono, Belinda Arunarwati, Peter V. Potapov, Svetlana Turubanova, Fred Stolle, and Matthew C. Hansen. 2014. "Primary forest cover loss in Indonesia over 2000–2012." *Nature Climate Change* 4, no. 8:730.

Petrenko, Chelsea, Julia Paltseva, and Stephanie Searle. 2016. *Ecological Impacts of Palm Oil Expansion in Indonesia.* Washington, DC: International Council on Clean Transportation.

UNORCID (United Nations Office for REDD Coordination in Indonesia). 2015. *Forest Ecosystem Valuation Study: Indonesia.* https://unredd.net/documents/global-programme -191/redd-and-the-green-economy-1294/forest-ecosystem-valuation-and-eco nomics/14398-forest-ecosystem-valuation-study-indonesia.html.

Forest Carbon Partnership Facility. 2016. *Carbon Fund Methodological Framework.* https:// www.forestcarbonpartnership.org/carbon-fund-methodological-framework.

———. 2017. *Technical Assessment of the Final ER-PD of Costa Rica.* https://www.forestcarbon partnership.org/sites/fcp/files/2017/July/TAP%20Report%20version%20post-CF14-%20 apr%2025%202017.pdf

Saatchi, Susan S., R. A. Houghton, R. C. Dos Santos Alvala, João Vianei Soares, and Yifan Yu. 2007. "Distribution of aboveground live biomass in the Amazon basin." *Global Change Biology* 13, no. 4:816–37.

Viana, Virgilio Mauricio, Andre Rodrigues Aquino, Thais Megid Pinto, Luiza M. T. Lima, Anne Martinet, François Busson, and Jean-Marie Samyn. 2012. *REDD+ and Community Forestry: Lessons Learned from an Exchange of Brazilian Experiences with Africa.* Washington, DC: World Bank.

Voigt, Christina, and Felipe Ferreira. 2015. "The Warsaw Framework for REDD+: Implications for national implementation and access to results-based finance." *Carbon & Climate Law Review* 9, no. 2:113–29.

PART III

Successful Experience in Inclusive Green Growth around the World

CHAPTER 12

China: Designing Policies to Enhance Ecosystem Services

Zhiyun Ouyang, Changsu Song, Christina Wong, Gretchen C. Daily, Jianguo Liu, James Salzman, Lingqiao Kong, Hua Zheng, Cong Li

To address the severe environmental crisis, policymakers in China are constructing a new governance strategy with major reforms across all social sectors to better balance development with ecological protection. It seeks to promote environmental quality and human livelihoods by enhancing and sustaining ecosystem services. The first step is the national ecosystem survey and assessment. The second step maps the services, identifying the crucial areas for ecosystem service provision. The third step addresses how best to secure ecosystem services and evaluate the effectiveness of their provision. And the last step is how to translate this into practical and effective policies, such as ecological functional zoning, ecological compensation, ecological restoration, and Gross Ecosystem Product (GEP) accounting. There are four key lessons that can be drawn from China's efforts to enhance green growth: match the ecological problem orientation with ecosystem service science, establish the sustainable supply of ecosystem services as a national goal, mainstream ecosystem services through policy innovation and financial mechanisms, and require new policy mechanisms to engage local residents and other stakeholders in conservation policymaking and implementation.

Decades of double-digit economic growth make China the fastest expanding major economy in history while saddling the country with likely the most severe environmental crisis faced by any civilization. China's ecosystems are quite fragile due to severe land degradation, erosion, desertification, water scarcity, and pollution. Ecological threats continue to grow in scale and severity across

China because of rapid urbanization and increased consumption of natural resources (Bryan et al. 2018). Wildlife habitat has declined, causing substantial losses in biodiversity, and poor air and water quality are causing human health problems. Political recognition of China's crisis started in 1998 when deforestation and erosion caused massive flooding along the Yangtze River. The floods killed thousands of people, made over 13.2 million people homeless, and cost US$36 billion in property damage.

To address this crisis, the Chinese government has realized it must change China's development model from unbounded growth to respecting environmental limits. President Xi and China's State Council are envisioning a new pathway forward, known as the creation of an *Ecological Civilization*. The aim is to improve livelihoods by achieving "harmony between humanity and nature." The Ecological Civilization is not simply a philosophical vision of social development. Policymakers are constructing a new governance strategy, with major reforms across all social sectors to better balance development with ecological protection. The Ecological Civilization captures China's approach to inclusive green growth. It seeks to promote environmental quality and human livelihoods by enhancing and sustaining ecosystem services.

Developing new policy mechanisms to improve environmental governance requires a strong scientific foundation. The first step is the nationwide ecosystem survey and assessment. The second step maps the services, identifying the crucial areas for ecosystem service provision and exactly where protection is needed. The third step addresses how best to secure ecosystem services and evaluate the effectiveness of their provision. And the last step is how to translate this into practical and effective policies.

In this chapter, we will describe the scientific initiatives surveying ecosystem services and mapping their provision, as well as their application in conservation policies such as ecological functional zoning, ecological compensation, ecological restoration, and Gross Ecosystem Product (GEP) accounting (fig. 12.1). We conclude with lessons learned on how to advance conservation based on ecosystem services.

Ecosystem Survey and Assessment

China's Ecosystem Assessment (2000–2010) provides the most detailed analysis of China's ecosystem services. The assessment covered China's major terrestrial ecosystems shown in figure 12.2. Led by China's Ministry of Ecology and

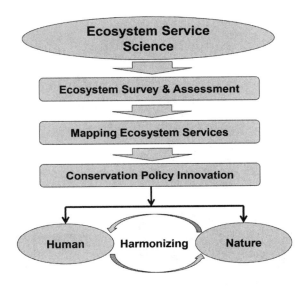

Figure 12.1. The framework of mainstreaming ecosystem services into policymaking in China.

Environment (MEE) and the Chinese Academy of Sciences (CAS), the assessment involved 3,000 researchers from over 139 institutions. It used a combination of geographic and ecological techniques to determine the current status and trends in ecosystem patterns, ecosystem quality, ecosystem services, and ecological problems. Data were compiled from multiple sources: 20,355 satellite images; biophysical measures (e.g., digital elevation model [DEM], soils, hydrology, and meteorology); 114,500 field surveys, biodiversity records; and special assessments from government agencies (e.g., surveys of desertification and of soil erosion).

The results showed, first, the basic terrestrial ecosystem assets of China (fig. 12.2): grassland occupied 2,836,758 km² (30.0 percent) of China's land surface, followed by forest (20.2 percent), agricultural land (19.2 percent), and deserts (13.5 percent) in 2010. Shrublands, wetlands, urban areas and others constituted the remaining 17.1 percent. Second, the assessment revealed that between 2000 and 2010, 195,803 km² (2.1 percent) of China underwent a change in major ecosystem types. Urban areas increased by 55,004 km², while agricultural land decreased by 48,234 km². Some 41,330 km² of forest, 9,111 km² of shrubs, and 21,103 km² of grassland were converted from other ecosystems—mostly (54.6 percent) from agricultural land. In addition to these major changes, wetlands also expanded mainly in the Tibet Plateau, the Three Gorges

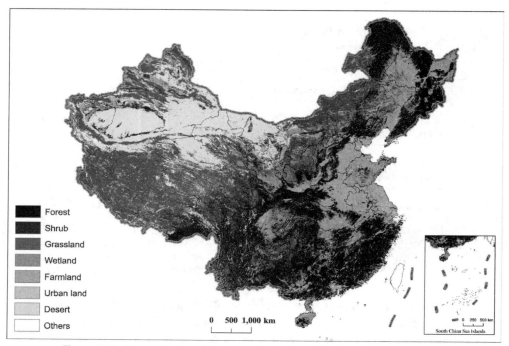

Figure 12.2. Ecosystem spatial patterns across China.

Reservoir Area, and the lower reaches of the Yangtze River and Northeast Plain. A total 18,569 km² of wetland were converted from other ecosystems: 38.2 percent from agricultural land and 27.5 percent from grassland.

Mapping Ecosystem Services

The ecosystem services of interest were (1) food production, (2) carbon sequestration, (3) soil retention, (4) sandstorm prevention, (5) water retention, (6) flood mitigation, and (7) provision of habitat for biodiversity (Ouyang et al. 2016). Data on food production was measured as crops converted to kilocalories (kcal) (fig. 12.3A). All of the ecosystem services (apart from food production) were modeled using InVEST (software models for Integrated Valuation of Ecosystem Services and Tradeoffs) and other biophysical models.

Food production in China is concentrated in the eastern plains (includ-

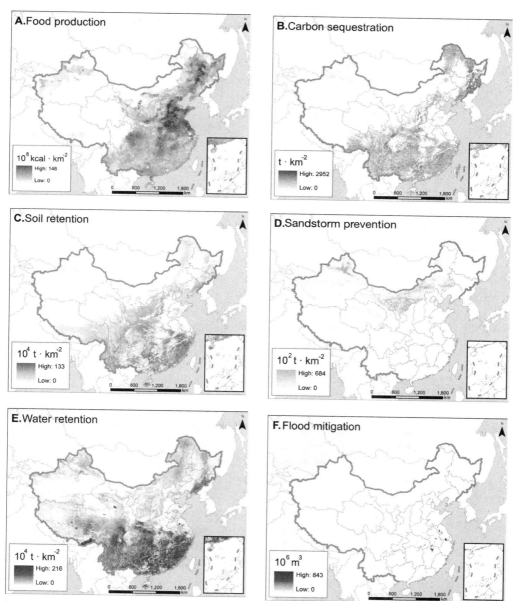

Figure 12.3. China ecosystem services spatial patterns: (*A*) food production (10⁸ kcal · km⁻²); (*B*) carbon sequestration (t km⁻²); (*C*) soil retention (10⁴ t · km⁻²); (*D*) sandstorm prevention (10² t · km⁻²); (*E*) water retention (10⁴ t · km⁻²); (*F*) flood mitigation (10⁶ · m³); (continued next page)

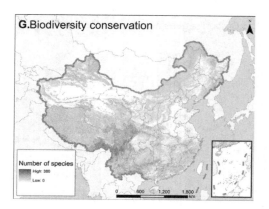

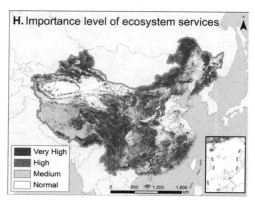

Figure 12.3. (continued) (*G*) provision of habitat for biodiversity (total species richness of endemic, endangered, and nationally protected species per county); and (*H*) index of relative importance of ecosystem services.

ing the Northeast Plain, North China Plain, and Middle and Lower Yangtze Plain) and the Sichuan Basin (fig. 12.3A). Important areas providing ecosystem services occur throughout the country (fig. 12.3B–G). The ecosystem service maps identify key hotspots for ecosystem services provisioning. This can help determine priority ecological areas for spatial planning (fig. 12.3H). Nationally, the priority areas are providing approximately 83 percent of China's carbon sequestration services, 78 percent of soil retention services, 59 percent of sandstorm prevention services, 80 percent of water retention services, and 56 percent of natural habitat for biodiversity, although they make up only 37 percent of China's land area.

China's Ecosystem Assessment provides the most recent and comprehensive analysis of the effectiveness of environmental protection policy mechanisms. In the past decade, the total change in ecosystem area is approximately 2 percent of China's total land area. The largest increases are in urban areas and forests, and largest decreases are in agricultural lands. We found that six ecosystem services increased between 2000 and 2010. Food production had the largest increase (38 percent) followed by carbon sequestration (23 percent), soil retention (13 percent), flood mitigation (13 percent), sandstorm prevention (6 percent), and water retention (4 percent), whereas habitat provision for biodiversity decreased (–3 percent) (fig. 12.4).

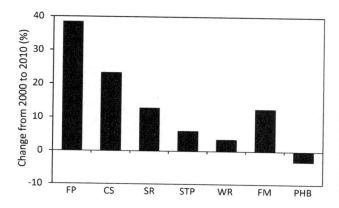

Figure 12.4. Ecosystem service changes during 2000–2010 in China: food production (FP), carbon sequestrations (CS), soil retention (SR), sand storm prevention (STP), water retention (WR), flood mitigation (FM), and habitat provision for biodiversity (PHB).

Conservation Policy Innovations

Based on the results from the national ecosystem survey and ecosystem services mapping, particularly of the spatial pattern of importance of ecosystem services, a series of new policies were designed and implemented. These include ecological functional zoning, ecological compensation, ecological restoration, and Gross Ecosystem Product (GEP) accounting.

National Ecological Function Zoning

Based on the data produced by mapping ecosystem services of China, in 2008, the MEE and CAS released the *National Ecological Function Zoning* (NEFZ), which was compiled over four years across fourteen government departments. In 2015, the MEE and CAS revised the *NEFZ* on the basis of the China Ecosystem Assessment. Sixty-three key EFZs (KEFZs) (fig. 12.5) selected from *NEFZ* were identified as crucial areas to ensure provision of ecosystem services. The KEFZs were focused on five ecosystem services: (1) water retention, (2) biodiversity protection, (3) soil retention, (4) sandstorm prevention, (5) flood mitigation (table 12.1; fig. 12.5). In total, KEFZs now cover approximately 49.4 percent of China's land area (4.74 million km²), providing approximately 78 percent of China's carbon sequestration services, 75 percent of soil conservation services, 61 percent of sandstorm prevention services, 61 percent of water resource conservation services, 60 percent of flood mitigation services, and 68 percent of natural habitat for biodiversity. These ecosystems represent important watersheds, forests, grasslands, and species habitat.

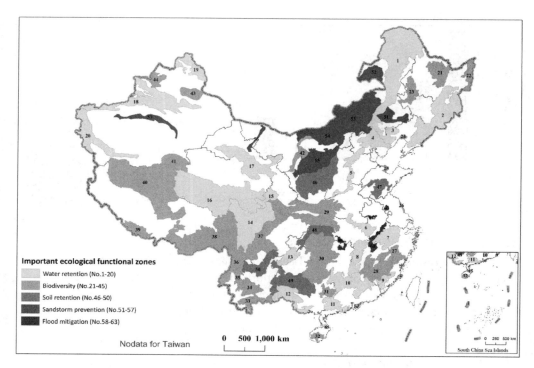

Figure 12.5. Key ecological function zones selected by China's Ministry of Ecology and Environment, grouped by ecosystem service.

The central and local governments are using KEFZs to determine the priority areas for conservation to protect crucial ecosystem services from urbanization/industrialization and agricultural development (Johnson 2017). For instance, the State Council formulated the Major Function Oriented Zoning–National Regional Development Strategy of China to guide spatial planning toward a sustainable pattern of development across China based on the KEFZs. The KEFZs areas were established as "restricted development zones," protecting the sustainable provision of ecosystem services from development.

The United Nations Environment Programme (2016) has described Major Function Oriented Zoning as a core innovation in China's new governance approach. For the first time, a major economy has designated "main functional areas" to "manage spatial use in accordance with the major ecological conditions of different localities."

Table 12.1. Key ecological function zones (KEFZs)

Functions	Number of zones	Area (x10³km²)
Water retention	20	2,035.6
Biodiversity protection	25	1,743.1
Soil retention	5	393.1
Sandstorm prevention	7	530.5
Flood mitigation	6	38.1

Ecological Compensation

The majority of KEFZs are located in remote rural and mountainous areas with high poverty rates. In these areas, local people depend heavily on natural resources and have limited employment opportunities. Restrictions on industry and agriculture will significantly impact these conventional livelihoods. Therefore, the Chinese government has implemented a series of ecological compensation policies to promote conservation and help communities transition toward new livelihoods, including a natural forest conservation program, ecological transfer payments, wetland ecological compensation, grassland ecological compensation, local ecological cooperation for water resource, and ecological forest compensation (Jin et al. 2016). Below, we briefly introduce ecological transfer payments and local ecological cooperation initiatives for water resources between Beijing and Hebei Province (Zheng et al. 2016).

Ecological Transfer Payments

In order to promote conservation in KEFZs, the central government began experimenting with ecological transfer payments in 2008, starting with 6 billion RMB (US$904 million) distributed across 230 counties. The number of participating counties and financial investments have grown every year. By 2017, the budget had grown tenfold, to 62.7 billion RMB (US$9 billion) benefiting 700 counties. To date, the central government has spent over 300 billion RMB (US$43 billion) on ecological transfer payments. The funding level is determined at the county level, taking into account population size, ecosystem types, spatial scale of KEFZs, natural land ratios, mean income levels, and so forth. The central government sums the total costs across the counties and cities in the given province. The Ministry of Finance transfers the funds to the provincial finance department who, in accordance with local conditions, formulates a transfer payment method to the municipalities and counties in KEFZs. The provincial government is responsible for effective fund allocation and supervision of activities. The central government regularly assesses the distribution and use of payments to monitor the effectiveness of fund transfers.

The funds are used to promote sustainable social and economic development by supporting two major activities: the enhancement of ecological restoration protection, and basic public services (e.g., education and healthcare). The allocation of funds is determined by the central government, which distributes the funds to provinces containing KEFZs. Provinces subsequently distribute the funds to respective counties within KEFZ boundaries. Local governments use the funds to curb ecological degradation by (1) compensating communities for forgoing industrial and development activities; (2) supporting national nature reserves and national park planning; (3) funding ecological restoration projects; (4) funding the recruitment and salaries of forest rangers to protect KEFZs; and (5) enacting pollution reduction and mitigation measures. Depending on the development condition of the counties, the funds can also be used to enhance basic public services, such as access to public education for school-age children and improving medical services.

The central government also regularly monitors local government performance in terms of fiscal responsibility, ecosystem services, water quality, public services, and poverty alleviation efforts. This determines whether payments will be reduced or enhanced. In regions where ecosystem services and quality continue to deteriorate, then 20 percent of the transfer payment is suspended until they are improved. For counties where ecosystems deteriorate for three consecutive years, the transfer payments are suspended for the following year. Payments do not resume until ecosystem services and water quality are restored to the pre-2009 level.

Paddy Land to Dry Land Program, Miyun Reservoir

Beijing is one of the most water scarce cities in the world. Beijing faces a water crisis requiring urgent solutions, and ecological compensation is a key strategy for protecting its limited water resources. The only surface water source for domestic water in Beijing is the Miyun Reservoir (fig. 12.6), and maintaining a sustainable and clean water supply is a major concern. Inflows into Miyun Reservoir have decreased due to upstream water withdrawals for agriculture and reduced precipitation. The mean annual inflow was 1.3 billion m^3 (BCM) in the 1960s, but fell to less than 0.4 BCM in the 2000s (a 70 percent reduction). Furthermore, nonpoint-source pollutants, mainly from agriculture, are harming water quality in the reservoir. The decreased river discharge and increased

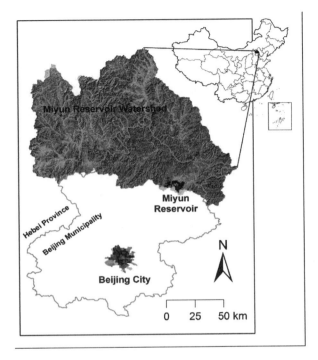

Figure 12.6. Miyun Reservoir Watershed is the main water source for Beijing Municipality, located about 100 km north of the city. Miyun Reservoir is the biggest artificial lake in Asia, spanning 188 km² with a storage capacity of 43.17×10^8 m³. About 4/5 of the watershed is located in Hebei Province with the remaining 1/5 in the greater municipality of Beijing. Beijing Municipality is the area outlined, and the areas of the watershed beyond Beijing are part of Hebei Province.

nutrient concentration have led to serious concerns over water pollution. If not addressed, this threat could eliminate the last local surface water source for Beijing.

The Miyun Reservoir supplies over half of Beijing's drinking water, but competition between the upstream Hebei Province and downstream Beijing Municipality for water resources is intensifying. Farmer incomes upstream are about one-third of farmer incomes downstream. While water demand by upstream townships is increasing for agricultural and industrial activities, Beijing's demand for water continues to increase due to its constant population growth (currently approximately 21 million residents) and urbanization. One of the greatest challenges in the watershed is addressing the interests of both upstream and downstream stakeholders to achieve socioeconomic goals in a sustainable manner.

Since 2001, Beijing and Hebei Province have jointly initiated the Paddy Land to Dry Land program (PLDL). The program's twin objectives are to increase water yield and reduce nutrient pollution. Rice cultivation through flood-irrigated paddies has been one of the primary causes of decreased water yield and high nutrient loads in the Miyun Reservoir. PLDL seeks to pay farmers to convert rice to dryland cultivation in order to enhance water provisioning and water purification services. In 2006, Beijing signed a "rice-to-dryland conversion" agreement with Chengde and Zhangjiakou Municipalities in Hebei Province. Beijing agreed to pay an average of 450 RMB per mu (~US$844 per ha in 2006; 15 mu = 1 ha and 8 RMB = US$1) in 2006 per year for land that was converted from rice to dryland cultivation, with payments adjusted to reflect market land-use values. In 2008, the Beijing government increased compensation to 550 RMB per mu (~US$1173 per ha in 2008) per year to ensure participation would not reduce household incomes. In practice the main land-use conversion was farmers switching from growing rice to growing corn. By 2010, upstream households in the Miyun Reservoir Watershed had converted all rice fields to dryland crops (total area = 103,000 mu).

The PLDL program has improved water provisioning and water purification services. The program increased water yield by $1.82*10^7$ m^3 per year and reduced total nitrogen (TN) and total phosphorus (TP) by 10.36 tons per year and 4.34 tons per year, respectively. The increase in water yield was 5 percent of the average runoff in Miyun Reservoir between 2000 and 2009. The net income for planting rice and corn was 8,602 RMB per ha (~US$1,036 per ha) and 1,501 RMB per ha (~US$221 per ha), respectively. Thus the program has not only increased water resources but also reduced the costs of treating water for TN and TP pollution. The economic benefits from improved water quantity were 12,341 RMB per ha (~US$1,763 per ha) and the reduced costs for TN and TP treatment were 46 RMB per ha (~US$6.5 per ha). In aggregate, these provide a positive cost: benefit ratio, with benefits exceeding by 1.3 times the cost of the PLDL program.

Ecological Restoration

China's degraded ecosystems now dominate the national landscape. Hence the Chinese government has been trying to restore or ecologically engineer degraded systems to enhance ecosystem services (Zhang et al. 2016). The central government has created a wide range of national and regional restoration programs: Grain to Green Program, Sanjiangyuan Nature Reserve in Qinghai Province, Beijing-Tianjin Sandstorm Control Program, Three-North Shelterbelt, Eco-environmental Protection and Comprehensive Management

Program of Qilian Mountains, Yangtze River Shelter Forests, Eco-environmental Protection and Comprehensive Management Program of Qinghai Lake, Integrated Management of Rocky Desertification in Karst Regions, and so forth. While some of these programs also receive ecological compensation as discussed above, the main goal of these programs has been to restore degraded ecosystems. This section focuses on the Grain to Green Program (GTGP) since it is one of China's largest and best known restoration programs.

Grain to Green Program

China has experienced widespread soil erosion for decades, which has led to serious social impacts and economic losses. Deforestation in the upper and middle reaches of the Yangtze and Yellow Rivers led to soil erosion, which made the system more vulnerable to massive floods.

The GTGP is a cropland conversion program that provides farmers with grain and cash subsidies to convert cropland on steep slopes to forest and grassland (Liu et al. 2008). From 2000 to 2013, the central government invested over 354.2 billion RMB (US$55.5 billion) in afforestation of the Yangtze River Basin (fig. 12.7), resulting in 31.8 million ha of new forests. The state pays farmers to return farmland to forests by subsidizing living expenses as well as grains and seedlings. The government gives farmers 2,250 and 1,500 kg of grain (or 3,150 and 2,100 yuan (~US$450 and 300) at 1.4 yuan per kg (~US$0.2 per kg) of grain) per hectare of converted cropland per year. Farmers are also given 300 yuan per hectare (~US$43 per ha) per year for miscellaneous expenses and a one-time subsidy of 750 yuan per hectare (~US$107 per ha) for seeds or seedlings. The timespan of the subsidies depends on the type of cropland conversion: two years if cropland is converted to grassland; five years if cropland is converted to economic forests using fruit trees; eight years if cropland is converted to ecological forests using tree species like pine and black locust. Payments are made only after the trees have been planted with a survival rate of 85 percent or more. In practice, this standard is flexible given local conditions. No taxes are collected on the converted cropland areas. To date, over 32 million rural households and 124 million farmers in 2,279 counties have participated in the program, making it one of the largest Payment for Ecosystem Services programs in the world.

The objectives of the GTGP are to (1) reduce soil erosion while (2) alleviating poverty by diversifying people's livelihoods. The GTGP has a hybrid gover-

Figure 12.7. Distribution of the Grain to Green Program (GTGP) in China.

nance structure using a top-down approach to set targets and a decentralized implementation led by provinces and local governments where household enrollment is voluntary. Farmers are paid to provide and improve ecosystem services. The aim is to offset opportunity costs associated with retiring land from cultivation and restoring them to either forests or grasslands. The National Development and Reform Commission (NDRC) oversees the planning of the program using household applications. The program is implemented by the National Forestry and Grassland Administration (NFGA), and the finances are managed by the Ministry of Finance. The NFGA sets out the national and provincial reforestation actions then distributes the land quotas to provincial governments. Provincial governments allocate the targets to counties and townships who, in turn, subsequently divide them among participating households. Local governments are held responsible for meeting the NFGA targets by signing liability agreements.

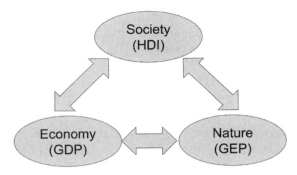

Figure 12.8. China is developing a national accounting system on the goods and services provided by ecosystems known as Gross Ecosystem Product (GEP).

The GTGP has enhanced the provision of water conservation and soil conversation. For example, in Hunan Province, from 2000 to 2005, scientists estimate that soil erosion declined by 30 percent, and surface runoff dropped by approximately 20 percent. The GTGP has reduced siltation by 22 percent in the Yangtze and Yellow River Basins. The program has also improved the physical properties of soil structure and soil nutrients by stabilizing top soils. In the Chaigou Watershed of Wuqi County of Shaanxi Province, the plots in the GTGP held on average 48 percent greater soil moisture than plots not in the program. The program also improved water conservation and reduced desertification. Increased vegetation in Minqin County in Gansu Province, for example, led to a 30 to 50 percent reduction of wind speeds on the soil surface, 15 to 25 percent increase in humidity, and reduction of dust in the air. Overall, the GTGP has significantly improved both forest and grassland quality over the past decade in China.

Gross Ecosystem Product Accounting

For decades Chinese officials have been evaluated for promotion in terms of their performance related to GDP. This fueled China's unprecedented economic growth rates but provided no incentive for conservation of ecosystem services. In order to align institutional behavior with ecosystem protection, the Chinese government is developing a balance sheet on ecosystem goods and services known as GEP accounting to evaluate the effectiveness and progress of conservation efforts and policy. GEP is defined as the monetary value of final ecosystem goods and services benefiting people, where value of a good or service is its price times its biophysical quantity. GEP, like GDP, is an accounting rather than an economic welfare measure (fig. 12.8).

Countries have adopted different indices to track macro level progress on human development (e.g., Human Development Index) and economic development (e.g., Gross Domestic Product), but there is a lack of a comparable index for the environment. GDP and GEP are parallel indicators to measure the performance of economic systems and ecosystems respectively. Both seek to identify comprehensive metrics for the flows they represent. GEP adopts the Millennium Ecosystem Assessment framework for categorizing ecosystem services, including as many ecosystem services as practical in order to ensure that it can be implemented and used by decision makers. The measure of GEP is meant to be a practical improvement over the status quo rather than a perfect measure. GEP measurement can improve through time with more data and understanding, much as GDP continued to be improved upon for decades after its first introduction.

The first step in constructing a measure of GEP is to assemble biophysical data defining metrics of ecosystem services (e.g., amount of grain production, water quality metrics and carbon sequestration). The second step is to find analogs for prices for ecosystem services. Some services are traded in markets (e.g., agricultural crops, timber, fish) and therefore market prices are readily accessible. For a few services, government policies have created markets, such as for carbon that have in turn generated prices. However, many ecosystem services are provided entirely outside markets and therefore lack prices. In some of these cases, prices can be represented through a variety of nonmarket valuation techniques. In many other cases, simple cost-based measures can be used. For example, ecosystem services such as water purification can be priced by looking at the cost of removing nutrients via water treatments plants. Similarly, flood mitigation services can be priced by evaluating the reduction in damages from reduced flooding. The value of a particular ecosystem service is simply its price in a particular location multiplied by the biophysical quantity of the service. GEP is then found by summing up the value over all ecosystem services. This approach provides a common unit comparator with GDP that uses tractable techniques, and readily available data.

Development of GEP has relied strongly on partnerships, such as with the International Union for Conservation of Nature (IUCN) to refine the definition of GEP to generate a pilot methodology for linking GEP to GDP. Since 2014, GEP has been supported by the Chinese government and multilateral organizations such as the Asian Development Bank (Ouyang et al. 2017). The pilot sites include four provinces (Qinghai, Hainan, Inner-Mongolia, and Guizhou); ten cities (Shenzhen, Lishui, Fuzhou, Tonghua, Qiandongnan, Hinggan League, Ganzi, Haikou, Puer, and Erdos); and more than one hundred counties (for instance, Arxan, Xishui, Shunde, Pingbian, and Eshan).

Conclusions

There are four key lessons to be drawn from China's efforts to enhance green growth: match the ecological problem orientation with ecosystem service science, establish the sustainable supply of ecosystem services as a national goal, mainstream ecosystem services through policy innovation and financial mechanisms, and management demand for policy innovation and financial mechanisms.

1. *Match the ecological problem orientation with ecosystem service science.* Since the 1990s, China has faced great challenges from water resource shortage, wildlife habitat loss, and increased disasters such as flooding and sandstorms from ecosystem degradation. Over this period, the concept and evaluation methods of ecosystem services have been developed. These provided a powerful instrument to help China to identify the crucial areas that must be protected. The science of identifying and measuring ecosystem services has become essential to conservation policymaking and innovation.

2. *Establish the sustainable supply of ecosystem services as a national goal.* Since 2005, China has made it clear that the purpose of conservation is to ensure the sustainable supply of ecosystem services. It has prioritized ecosystem service supply, particularly water retention, soil retention, sandstorm prevention, flood mitigation, and endangered wildlife habitat protection as national goals. At the same time, it has required that resource exploitation and land development cannot take place at the expense of declining ecosystem services.

3. *Mainstream ecosystem services through policy innovation and financial mechanisms.* China first carried out a comprehensive national ecosystem survey and assessment. It then mapped ecosystem services, and identified the crucial areas for ecosystem services as national KEFZs. Finally, based on the results from the earlier steps, China developed and implemented a series of polices to ensure conservation of ecosystem services. These included ecological compensation and financial transfer payments, eco-

system restoration projects, regional ecological cooperation, and a new evaluation mechanism (GEP accounting) for government performance on conservation based on output of ecosystem services.

4. *Management demand for policy innovation and financial mechanisms.* China still faces huge conservation challenges. In many parts of China, ecosystem quality and ecosystem functions remain low. The problems of soil and water loss, desertification, ecological deterioration of watersheds, and the reduction of natural habitats (e.g., coastal ecosystems) have not stabilized and continue to be prominent. Conservation policies are needed to help more local residents and provide poverty alleviation. In order to ensure effective policy implementation, it is essential to engage local residents and other stakeholders in conservation policymaking and implementation.

Key References

Bryan, Brett A., Lei Gao, Yanqiong Ye, Xiufeng Sun, Jeffery D. Connor, Neville D. Crossman, Mark Stafford-Smith et al. 2018. "China's response to a national land-system sustainability emergency." *Nature* 559, no. 7713:193–204.

Jin, L., et al. 2016. *Advances of China Eco-Compensation Policies and Practices in All Sectors.* Beijing: Economic Science Press.

Johnson, Christopher N., Andrew Balmford, Barry W. Brook, Jessie C. Buettel, Mauro Galetti, Lei Guangchun, and Janet M. Wilmshurst. 2017. "Biodiversity losses and conservation responses in the Anthropocene." *Science* 356, no. 6335:270–75.

Liu, Jianguo, Shuxin Li, Zhiyun Ouyang, Christine Tam, and Xiaodong Chen. 2008. "Ecological and socioeconomic effects of China's policies for ecosystem services." *Proceedings of the National Academy of Sciences* 105, no. 28:9477–82.

Ministry of Ecology and Environment of the People's Republic of China. 2015. http://www.mee.gov.cn/gkml/hbb/bgg/201511/t20151126_317777.htm.

Ouyang, Zhiyun, et al. 2017. *Developing Gross Ecosystem Product and Ecological Asset Accounting for Eco-compensation.* Beijing: Economic Science Press. Ouyang, Zhiyun, Hua Zheng, Yi Xiao, Stephen Polasky, Jianguo Liu, Weihua Xu, Qiao Wang et al. 2016. "Improvements in ecosystem services from investments in natural capital." *Science* 352, no. 6292:1455–59.

United Nations Environment Programme (UNEP). 2016. *Green Is Gold: The Strategy and Actions of China's Ecological Civilization.* Geneva, Switzerland: UNEP.

Zhang, Ke, John A. Dearing, Shilu L. Tong, and Terry P. Hughes. 2016. "China's degraded environment enters a new normal." *Trends in Ecology & Evolution* 31, no. 3:175–77.

Zheng, Hua, Yifeng Li, Brian E. Robinson, Gang Liu, Dongchun Ma, Fengchun Wang, Fei Lu, Zhiyun Ouyang, and Gretchen C. Daily. 2016. "Using ecosystem service trade-offs to inform water conservation policies and management practices." *Frontiers in Ecology and the Environment* 14, no. 10:527–32.

CHAPTER 13

Costa Rica: Bringing Natural Capital Values into the Mainstream

Alvaro Umaña Quesada

Costa Rica made headlines for its record-high deforestation rate, but it is now widely known both for its astounding biodiversity and for its long-standing efforts in conservation. Since the early 1990s, the country has developed a strong system of national parks and protected areas, many of which are World Heritage Sites. Costa Rica devoted considerable financial resources, both internal and external, to conservation and forest management, using many different financial incentives and mechanisms. Costa Rica is notable for using debt-for-nature swaps more extensively than any other country, with both commercial and bilateral debt. Other key mechanisms pioneered in concept and implementation are tax deductions to incentivize reforestation, bans on deforestation, and the famous payment for ecosystem services (PES) program at the country level. In sum, Costa Rica reversed its deforestation and is now undergoing net reforestation, along with many improvements in economic, health, education, and other dimensions of well-being. In total, 27 percent of the country's land area is under protected status, and 20 percent (>1 million ha) has at some time been under PES programs, incentivizing conservation and restoration for the realization of a wide array of benefits from nature. These policy and finance mechanisms evolved considerably through time. Some produced disappointing results and were redesigned or discontinued entirely, but they constituted crucial steps in the cultural evolution accompanying the advances. Today, the effectiveness of the system is thanks to its many complementary and interacting influences in securing the environment and also land and livelihoods of rural and indigenous peoples. The global interest in the system traces back to the adaptability of many approaches to other countries.

....

Costa Rica is unique in several ways. A key factor is the country's geographic location as a narrow land bridge between the great landmasses of North and South America (fig. 13.1). Its rugged topography includes mountains that rise up to 3,800 meters above sea level, and its land bears two different seasonal regimes that contribute to its extremely varied habitats. Costa Rica's biological diversity is outstanding in the Americas as well as in the tropical world as a whole. With a land area of only 51,000 square kilometers, it contains nearly 5 percent of all plant and animal species known to exist on the planet.

Christopher Columbus landed in what is now Port Limon in 1502, on his fourth trip. Impressed by the abundance of gold gifts, he named the territory "Costa Rica." Although the country has some gold deposits, mining has been banned, and Costa Ricans have become aware that the real richness lies in its "green gold": its biological endowment. This change in values and attitudes did not come easy, and it has taken a long time to evolve.

In the early 1520s, when the first Spanish *Conquistadores* and settlers arrived, they found natural forests from coast to coast. Extensive fixed farmlands of indigenous tribes mixed with dry forest in the Nicoya Peninsula of the Northwest. In wetter areas, shifting agricultural systems were widespread; their common characteristic is that they allowed the forest to grow back.

The arrival of Spanish settlers brought new diseases that decimated the indigenous population, and as the Spanish culture slowly started to dominate most of Costa Rica, the agronomy of native cultures was abandoned. This process took place in most of the country, except the southern mountains of Talamanca (what is now Amistad National Park), where the Spanish were never able to conquer the native tribes.

Before the arrival of the Spanish, Costa Rica's population was estimated to be in the hundreds of thousands, most likely close to 500,000 people. Afterward, the impact of land theft and disease was so catastrophic that the indigenous population had shrunk to ~17,000 in 1569 and ~8,000 in 1801. Living primarily in the valleys and plateaus of the interior, this diminished population remained isolated and poor.

Toward the middle of the nineteenth century, the fertile Central Valley was almost completely occupied, and colonists moved into the forests. Waves of settlers and immigrants destroyed large tracts of the country's forests and started a culture and frontier mentality that persisted until recent decades among the rural population. There are two traditional myths about tropical forests: that the forest is the enemy of the peasant, and that lands where all

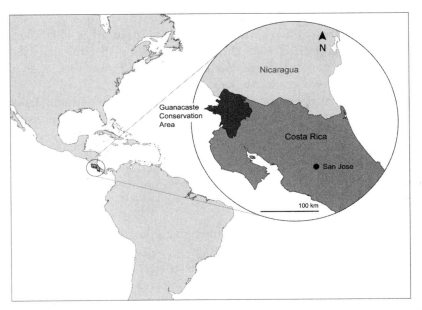

Figure 13.1. Costa Rica's importance for global biodiversity is far greater than its area would suggest—due to its location on the land bridge between North and South America, along with its rugged topography and climatic variability, The country hosts nearly 5 percent of the world's plant and animal species. The Guanacaste Conservation Area (*inset*) plays an especially important role, covering less than 2.5 percent of Costa Rica's land area but home to two-thirds of the country's biological diversity.

these lush forests grow must be productive. In addition, most tropical countries require "improvements" (i.e., removing the forest cover) in order to gain legal title over the land.

The development of coffee agriculture on the land's fertile volcanic soil had a major impact on land colonization, and by the mid-1850s, Costa Rica started exporting coffee to Europe. The penetration of these markets caused a major social transformation and initiated a process of integrating Costa Rica into the world economy. Traditional coffee plantations kept large canopy trees for shade and they planted nitrogen-fixing trees, so the overall impact of coffee was not as damaging as the new unshaded varieties and practices that replaced some of the old plantations. Today, however, many producers are going back to the traditional system, and they market their coffee through Fair Trade or organic markets.

Forest cover in Costa Rica has undergone dramatic changes, decreasing from close to 95 percent when the Spanish arrived to 70 percent in 1950, and to below 30 percent in the late 1980s. During the decades of 1940 through 1980, very rapid deforestation took place, reaching over 1 percent per year. This was one of highest deforestation rates in Latin America.

This massive transformation of the landscape resulted from a combination of domestic policies, as well as international markets and political pressure. The early period of deforestation saw forests rapidly converted into agricultural or cattle ranching areas, which benefited from generous land titling and cheap credit. Even the World Bank gave Costa Rica credit to expand cattle production in the 1970s. High international prices for beef and other expansive crops like coffee and bananas exacerbated these policies' acceleration of deforestation.

The development of a road network throughout the country also had a significant impact on deforestation. When any new roads were built, deforestation followed closely, and forests near the new roads were gone within five years. This happened, for example, in the Perez Zeledon area and adjacent valleys in the southwest when the Inter-American Highway was completed in the 1950s.

A critical problem was the fact that forest owners perceived little value in standing trees, and the legal and institutional structure reinforced this conception. Trees were not capital goods equivalent to cattle or tractors. One could go to the bank and borrow against these assets but trees were not acceptable as collateral. Only dead trees were valued, and only for their wood. Loggers focused on highly valued trees of precious woods and did considerable damage to the remaining forest to extract only these few trees per hectare.

This trend was stopped by a number of significant pressures that emerged in the 1980s. Political and economic instability created by wars in Central America, and the collapse of global meat, sugar, and coffee markets led to the abandonment of significant agricultural and cattle land. At the same time, Costa Rica had begun to develop conservation policies like the creation of national parks and protected areas. An emergent conservation movement started to call for change.

Costa Rica Begins to Develop Conservation Policies

The need to preserve key examples of Costa Rica's extraordinary natural resources began to be felt as early as the middle of the nineteenth century after settlers had colonized most of the Central Valley. Early decrees introduced the concept of inalienable areas—lands that could not be privately owned and should be preserved.

In 1913, the crater and lake of Poás Volcano were protected. In 1945 the words *national park* appeared for the first time in Costa Rica legislation, and this same law declared that two kilometers on each side of the future Pan-American Highway were to be protected. In 1955, a law created the Tourism Institute and gave it the right to establish and maintain national parks. At the same time, the law also declared that those areas within a radius of two kilometers from volcanic craters were also national parks.

These early measures were not well enforced, primarily owing to the lack of resources. For example, the oak forests near the Pan-American Highway vanished in a few years, mostly having been converted to charcoal by local residents. Nevertheless, the idea of protecting national parks continued to gain strength and in 1969 a new forestry law established a National Parks Department.

Efforts to create and manage a system of national parks and protected areas has received political support from all subsequent governments. The following year, the José Figueres administration (1970–1974) started declaring a number of national parks: the Cahuita, Santa Rosa, Poás, Irazú, Manuel Antonio, and Rincón de la Vieja national parks, as well as the Guayabo National Monument, the only archeological park in Costa Rica. The Daniel Oduber administration (1974–1978) gave greater independence to the National Park Service and added key new parks: Barra Honda, Chirripó, Corcovado, Tortuguero, and Braulio Carrillo, as well as the biological reserves Hitoy-Cerere, Carara, and Caño Island. These efforts were continued by the Rodrigo Carazo administration (1978–1982), which added Cocos Island, Palo Verde, and La Amistad national parks.

In 1978 two important events took place. First, the creation of the National Parks Foundation (NPF) by the government. Under Costa Rican law, foundations operate as private entities, with three members appointed by the founder, a representative of the government, and a representative of the municipality where the foundation is based. Since the government created the foundation, it had control over four out of five of its members. The NPF was the first of a number of foundations that have been created for conservation.

The second important event was the start of a fiscal incentive program for reforestation paid by the state through tax deductions. The loss of forest cover was the most important environmental issue at that time and the government decided to support reforestation. The program worked by allowing anybody who had a tax liability to buy a reforestation contract and take the tax deduction (50 percent in the first year and the other 50 percent over the next four years). Forest entrepreneurs were allowed to buy land, reforest it, and sell the tax deduction. In turn they would keep the land and the trees. Unfortunately, since the buyers were only interested in the tax break, and you could deduct half in the first year, many of the plantations failed.

The Luis Alberto Monge administration added several protected areas such as La Selva, which extended Braulio Carrillo National Park to the northern plains, and the Pacuare River, which protected a major whitewater attraction threatened by hydroelectric development. During this period, the US Agency for International Development (USAID) started to play an important role in the natural resources area, and the first project in the forest sector included subsidized credit to entice farmers to go into reforestation.

Through these sustained efforts, Costa Rica's system of national parks and reserves protects the nation's most outstanding examples of its natural and cultural diversity. The national parks and reserves, which account for over 6,000 square kilometers (12 percent of the country's land area), constitute the core of the system. An additional 15 percent of the territory falls under different categories of protection, such as forest reserves, protected zones, wildlife refuges, and indigenous people's reserves. This critical conservation effort spearheaded by the state is complementary to the policies related to payment for ecosystem services (PES) that would be developed later.

Creation of the Ministry of Environment and Energy

The creation of the Ministry of Natural Resources, Energy and Environment (MIRENEM) in 1986 during the administration of President Oscar Arias (1986–1990) was an important political landmark. National parks and all forestry and wildlife related agencies, which were originally under the Ministry of Agriculture, were consolidated under the new ministry that would later be named Ministry of Environment and Energy (MINAE). For the first time in Costa Rican history, environment had a cabinet-level position, and this was an important political triumph for conservation and for launching sustainable development. A new forestry law had been passed at the end of the previous administration, and it became effective as the new ministry took over.

The new administration also launched a participatory process to establish a national strategy for sustainable development, decentralized and regionalized the system of protected areas, developed new financial mechanisms for conservation that attracted new bilateral donors, initiated the National Biodiversity Institute, and spearheaded the Guanacaste National Park project, as well as creating Arenal National Park.

By the time the Arias administration took over in 1986, the program of fiscal incentives for reforestation had gotten out of hand. The Finance Ministry had complained that they could not know how much people would deduct until after they had presented their tax declarations, and the size of the program had gone into the billion colones (US$10–20 million) range.

At the same time, the new ministry became aware that supporting reforestation through fiscal incentives had been a mistake from the start. The priority should have been to protect the remaining forests, not to support wealthy individuals or corporations who were benefiting from the program. Yet, it was necessary to create a culture of reforestation and to develop policies that would give value to standing trees and forests.

A new program needed to be designed and the focus was small landholders and peasant organizations. The Finance Ministry continued to support the program but they would give a budgeted amount that was used to support organizations, like cooperatives or agricultural centers with members who would be the beneficiaries of a newly named CAF (a forestry certificate introduced by the new law): it would provide funding in advance so that peasants and farmers could plant trees. The maximum size allowable was 5 hectares, prioritizing small landholders.

This effort got a considerable boost when Costa Rica started to experiment with debt-for-nature swaps in 1988. This mechanism started to be implemented in the late 1980s with developing country debt held by commercial banks, which was then sold at a discount in secondary markets. In a typical transaction, a donor (private, nongovernmental organization [NGO], or developed country government) would purchase debt at a considerable discount, and then go to the Central Bank of the recipient government and exchange it for local currency bonds or future payments, also in local currency. The idea is that all parties are better off: the country reduces its debt, the projects receive local funding, and the donor magnifies its impact. And new protected areas are established.

After an initial wave of commercial debt-for-nature swaps, the mechanism was taken up by several donor countries that used it with their own bilateral debt. For example, the United States passed a Tropical Forestry Conservation Act that allows its debt to be swapped for conservation. Costa Rica became the country to benefit most from this act. Today, developing country debt is over US$1 trillion, so there is considerable potential for this mechanism to be used throughout the world.

Through this new mechanism, Costa Rica also attracted two new bilateral donors: Sweden and the Netherlands. They provided the funds for the two largest commercial debt-for-nature swaps. The Swedish debt-swap went to Guanacaste National Park, while the Dutch debt-swap was dedicated to create a fund to finance forest management and reforestation. This added considerably to the resources provided by the government. Donations and grants totaling approximately US$12 million cash were used to purchase Costa Rican commercial debt titles equivalent to US$75 million face value. In exchange for the titles, the Central Bank returned more than US$35 million in local currency bonds that were very liquid (short-term maturation period). Therefore the donors almost tripled the value of their contribution, while the government reduced its foreign debt.

Evolution of Forest Incentives to PES

Costa Rica experimented with a wide variety of financial instruments including soft loans, a fiscal incentive program, the CAF forestry certificates, as well as other measures like restricting forest permits and promoting sustainable forestry. In fact, in 1985 the Central Bank had approved a debt-for-equity swap for a company that was going to do sustainable forestry. Although a wide variety of incentives were implemented, they had limited success in stopping deforestation and encouraging reforestation. Nevertheless, these early incentives provided the experience and lessons, as well as paving the way for the creation of the program of payment for ecosystem services (PES).

The new Forestry Law 7575 was approved in 1996, which introduced new mechanisms and instruments. First, it banned conversion of forest lands, punishable by prison sentences rather than fines, effectively lowering the opportunity cost for forest conversion to pastures or agriculture. And second, it introduced payments for protecting the forests, managing forest sustainably, or undertaking reforestation projects, effectively creating the basis for the PES program.

The forestry and biodiversity laws use the term *environmental services*, and Forestry Law 7575 defines four different categories of services:

- Carbon sequestration, capture, and long-term storage of carbon dioxide.
- Hydrologic services, provision of clean, steady water flows, and protection of water catchment areas.
- Biodiversity conservation and sustainable use.
- Scenic beauty, aesthetic values.

Through the National Forestry Financing Institute (FONAFIFO), also created by Law 7575, PES became operational in 1997, and it is the primary institution created to manage the PES system.

The system is land-based and FONAFIFO signs five-year contracts with individual private forest owners, with a priority to those that hold clear land titles, agreeing either to maintain existing forests or undertake reforestation or sustainable forestry projects. Forest engineers monitor compliance on a yearly basis.

In exchange for the payments, the landowners transfer the "rights" to the ecosystem services to FONAFIFO, which acts as a wholesale intermediary for these services. Initially, it was expected that an international carbon market would develop quickly, and that the credits held by FONAFIFO could be sold in these international markets. However, this assumption turned out to be wrong, and Costa Rica has sold only US$2 million in carbon credits to Norway, in an effort to spark the private carbon market.

Structure and Design of PES: Institutional Architecture, Complementary Policies

The PES program started on a first-come, first-serve basis and the challenge was to get people to sign, so during the first phase there was no targeting. An important source of support came from a World Bank loan and GEF grant to develop and provide financing complementary to that provided by the government. The main source of support from the government came in the form of a 3.5 percent tax on fuels that is managed by FONAFIFO. There would be a follow-up loan with the World Bank, known as Ecomarkets II that had similar structure and provided advice on targeting biodiversity corridors and introduced a differential payment for water.

FONAFIFO relied on existing NGOs and cooperatives and regional agricultural centers to act as intermediaries to reach individual landowners. These intermediaries provided technical assistance and monitoring of the projects. The program was regionalized from the beginning, and FONAFIFO established seven offices throughout the country to be able to handle and decide on individual applications that would get funds from the Finance Ministry. In 2009 the Comptroller General decreed that FONAFIFO should be integrated directly into the government structure.

Most of the PES resources—over 80 percent—have gone to forest conservation and reforestation. The sustainable forestry option was not implemented after the Ministry of Environment and Energy (MINAE) decided that sustainable forestry should be profitable without incentives.

The PES program received widespread attention because it was, and still is, the only country-level program. The PES approach has been widely criticized and debated by researchers and academics. Some considered that it was redundant and did not have a clear focus; others argued that it had little value in stopping deforestation; while still others argued that it was necessary to keep forest standing due to the low level of law enforcement (Porras et al., 2013).

One important fact to bear in mind is that PES was not acting alone, but rather in concert with complementary policies. The most important was the creation of the system of national parks and protected areas, initially by decree and later ratified by congress. Costa Rica has invested hundreds of millions of dollars in the purchase of inholdings and the management of the park system, recognized as one of the best in the world.

A second complementary policy was the ban on land-use change from forestry to pasture and agricultural lands. Another had to do with the establishment of private conservation reserves of which there are more than 200 in Costa Rica. A shining example is the Monteverde reserves, which are all technically private and were started by US Quakers in 1954. The area received critical support from Swedish children and many others, were established with private contributions, and today are run by NGOs.

A third key complementary policy was the ban on sustainable forestry and the decision that educational, research, and tourism activities could be carried on in protected areas. It is difficult to weigh the relative importance of these measures, which also contributed to slowing and eventually reversing the deforestation process. We cannot isolate the impact of PES.

The PES program started as an experiment because a political window of opportunity opened when the Figueres administration (1994–1998) agreed to a 5 percent tax on fuels for PES (later the tax was lowered to 3.5 percent). Support of the Finance Ministry was also invaluable in exceeding its counterpart contribution to the World Bank loans. Although the tax cannot formally be considered a carbon tax, it is a good proxy. It has the additional advantage that the revenues go to forest protection and reforestation, and in that sense it is unique.

The PES has been the subject of considerable attention, analysis, and criticism. It was not a program designed from scratch by researchers. It was a political and institutional experiment that evolved with time, becoming much more focused since its inception over twenty years ago.

The reversal of tropical deforestation in Costa Rica is the product of many policies acting together, and the PES has played a key role, with political support of all governments since the early 1980s. This continuity has been essential. It is undoubtedly an achievement that has served as an example to other countries like Mexico and Ecuador. Over eighty countries have sent delegations to Costa Rica to study this effort.

Financing and Impact of the PES Program

The program has been financed primarily by the government through earmarked tax revenues (3.5 percent tax on fuels, equivalent to US$12–16 million per year depending on oil prices) and two World Bank loans totaling over US$80 million, as well as grant contributions from the GEF and KFW (German Cooperation Bank). Costa Rica is the only country to have financed such a program with loans.

There is also a water fee that started to be charged in 2006 when taxes were increased, with 25 percent of collections destined for PES in strategic water catchment areas, and contributions from the private sector. There is one municipal water company that has incorporated PES in its rate structure and actually charges a fee to customers. Some private hydroelectric projects also contribute, but only about 3 percent of the program is financed by private funds.

Today FONAFIFO has a staff of around ninety people, including the regional offices that continue to run the PES, and is a decentralized organ of the Ministry of Environment and Energy (MINAE).

It operates directly with the Finance Ministry, and has other sources of international funding and the capability to create other funds. This was achieved through the Ecomarkets II program, which included a clause creating an environmental bank foundation (FUNBAN) that can create funds like the Sustainable Biodiversity Fund.

The international carbon markets were an important target for FONAFIFO, and the initial US$2 million transaction with Norway led to "Certified Tradable Offsets" (CTOs), as the carbon credits were known at the time. These CTOs were offered in the Chicago Climate Exchange but never sold. The international carbon markets were a disappointment because the official markets failed to develop. The forestry issue became very complex in the climate negotiations, so official transactions were not possible, and the voluntary markets offered very low prices. In fact, Costa Rica's internal price, estimated near US$8 per ton of carbon dioxide equivalents (Porras et al. 2013), always exceeded the voluntary international carbon markets by almost 50 percent. A REDD+ market, which is consistent with what Costa Rica's PES can provide, has failed to materialize.

The PES program focuses on five types of private land use: (1) forest protection, (2) commercial reforestation, (3) agroforestry, (4) sustainable forest management, and (5) regeneration in degraded areas. As mentioned earlier, option 4 was not implemented. These types of land uses served as a proxy for providing the four ecosystem services denoted above. Water recharge areas were recognized, and a differential payment system introduced. For example, the basic payment is US$64/ hectare per year for a period of five years. Originally this payment was calculated based on the opportunity cost of cattle ranching. Properties located in conservation gaps receive US$75/hectare per year and those in water recharge areas receive US$80/hectare per year.

Today, the program has become more focused and applications are ranked on a point-based system. There are many more applicants than the program can handle and only three out of every ten applicants make it to the program. This is mostly due to legal and financial constraints. The yearly cost of the program has been US$15–20 million.

An important feature is that about 10 percent of the beneficiaries are indigenous communities that apply to the program as a community, not as individual landowners. During the last five years, this has resulted in a transfer of more than US$25 million to these communities, who use the funds to build schools or health centers. Other important beneficiaries have been conservation projects outside of formal protected areas, run by private funds such as on lands bought for conservation surrounding parks in the Guanacaste area. We will consider this case next.

The Sustainable Biodiversity Fund (SBF) is a key innovation, and it was created with grants from the GEF and later the German Cooperation Bank (KFW) at US$7.5 million each, as well as local contributions from green credits cards and compensation schemes for clean travel. The SBF was capitalized to a level of US$18 million, and in 2016 started making payments for biodiversity in hotspot areas. The SBF is attempting to work with specialized conservation groups, for example birdwatchers, to establish a mechanism whereby private contributors could support PES biodiversity payments in biological corridors or hotspots.

The twenty-year evaluation of PES highlights strong agreement on the positive nature of the program:

> The program has had positive impacts since its inception. Between 1997 and 2012, it has protected more than 860,000 hectares of forest, reforested 60,000 hectares and supported sustainable forest management in almost 30,000 hectares. More recently it promoted natural regeneration of almost 10,000 hectares. This totals nearly one million hectares under the PES scheme at one time or another, as well as 4.4 million trees planted under agroforestry systems since 2003. This is a substantial achievement for a developing country of just 51,100 square kilometers. By 2010, roughly 52 percent was under some sort of forest cover. (Porras et al. 2013)

Overall, Costa Rica has invested more than US$400 million in this program over the course of almost twenty years, purchasing ecosystem services from over 16,000 private providers, including indigenous communities (around 10 percent of the program). This is a highly significant achievement for a small developing country and is consistent with its comparable conservation investments.

The Guanacaste Conservation Area (ACG)

The Guanacaste Conservation Area (ACG) is one of the most ambitious ecological restoration efforts in the world. Encompassing 1,260 km^2 (almost 2.5 percent of Costa Rica's land area), it includes tropical dry forest, rain forest, cloud forests, and 430 km² of marine protected areas (fig. 13.1). It is home to 375,000 macrospecies, which represent two-thirds of Costa Rica's and 2.4 percent of Earth's biological diversity (Pringle 2017). It grew from the original Santa Rosa National Park, one of Costa Rica's first generation of national parks declared in 1971 by President José Figueres. For a detailed account of the fascinating history of Guanacaste National Park Project see Umaña (2016).

Tropical dry forest is one of the most endangered life zones in Mesoamerica, and the establishment of Guanacaste National Park Project (GNPP) was intended to protect the largest remaining tract of dry forest from California to Panama. The project needed to be large enough to maintain habitats for healthy populations of all animals and plants that are known to have originally occupied the region, and to contain enough duplicate habitats to allow intensive use of some areas by visitors and researchers. The effort included the restoration of large areas of species-rich and habitat-rich tropical dry forest, and the main mechanisms utilized were fire control by managers, grass control by cattle, and tree seed dispersal by wild and domestic animals. In addition to being a major cultural resource, the park has made significant scientific contributions, developed the concepts and practice of "para-taxonomists," and "bioliteracy," focusing on all schoolchildren that inhabit the conservation area.

Para-taxonomists are hybrids between cultures. On one hand, they are in the world of science, communicating and providing new data to global experts on taxonomy and ecology. On the other hand, they are members of their communities without official scientific credentials: peasants, school parents, churchgoers, and other citizens. Para-taxonomists are trained to collect specimens, and many acquire impressive scientific and practical knowledge: it is not unlikely that their sons and daughters will end up as professional biologists.

The second area in which the ACG has made important innovations is in their bioliteracy efforts, supported by GDFCF and the national school system in the area. This program started in 1987 with students from fourth, fifth, and sixth grades, teaching them the essentials of biodiversity conservation as well as how they can use it to improve their livelihoods.

Building on this basis, the ACG has benefited from the long-term intellectual and scientific leadership of biologists Daniel Janzen and Winnie Hallwachs. The original proposal for GNPP was for a total of 82,500 hectares, and a total cost, including land purchases, of approximately US$12 million. Today we can be proud to see that the land area target has been exceeded by a factor of two, and over five times in terms of resources donated.

The ACG received critical political and financial support from the Arias administration (1986–1990). Since then the ACG has utilized a variety of financial mechanisms and both public and private funds and contributions. A critical boost came from Sweden in 1988, which provided a US$3.5 million grant that was used to purchase US$24.5 million of Costa Rican commercial debt, for which the Central Bank provided US$17 million in liquid government bonds. This transaction was executed through the National Parks Foundation (NPF) and created a trust fund that has provided additional support to ACG throughout almost thirty years. This fund is almost entirely exhausted.

The ACG has also been the recipient of funds from the two bilateral debt-swaps Costa Rica has done with the United States under the Tropical Forestry Conservation Act (TFCA). Costa Rica obtained US$26 million for conservation from the first operation in 2007. This operation included a grant for land purchases in the ACG. The second, in 2010, for US$27 million to create the Costa Rica Forever Fund (CRFF), is considered next. These debt-swaps also leveraged private funds, US$2.5 million in the first transaction, and US$3.9 million in the second one. The ACG has also received funds from the CRFF.

Summarizing a rich history, Janzen estimates that, from 1985 to 2015, the "off-the-shelf" cost of ACG has been approximately US$107 million, excluding efforts related to the National Biodiversity Institute (INBío). The Costa Rican government made a contribution equivalent to US$17.5 million through the debt-swaps and then had to pay US$16 million for Hacienda Santa Elena, for a total of US$33.5 million. In turn, external fundraising efforts have yielded over US$65 million. This impressive figure is almost twice as much as the government contribution so, in effect, external donors matched Costa Rica's contribution on a two-to-one basis.

Even more remarkable is the fact that this effort has been the result of a partnership between the government, national and international NGOs, and key donors. In total, there have been more than 15,000 individual donors to this project, supporting this as a global restoration effort.

The Costa Rica Forever Association

The Costa Rica Forever Association is a recent conservation-focused institution. It administers the Costa Rica Forever Irrevocable Trust Fund and the second debt-for-nature swap between Costa Rica and the United States. Together these resources amount to approximately US$56 million.

The Forever Costa Rica program was developed by a public-private partnership with the Costa Rican government, The Nature Conservancy (TNC), the Gordon and Betty Moore Foundation and the Linden Trust for Conservation. This enabled the establishment of a permanent mechanism for resourcing its activities. At the same time, the Costa Rica Forever Association manages the resources and monitors the program's implementation.

The association has signed a five-year agreement with SINAC, the entity that manages the system of protected areas, to define priority areas for its work program. This strategy focuses on the following:

1. Closing the gaps in ecological representativeness
2. Increasing management effectiveness
3. Identifying and incorporating adaptation and mitigation activities related to biodiversity in the terrestrial and marine protected areas vulnerable to global climatic variability
4. Establishing a sustainable source of funding for existing and/or future protected areas

Costa Rica's conservation had traditionally prioritized terrestrial protected areas. Through the establishment of Costa Rica Forever, the country doubled the size of its marine protected areas. Efforts were also undertaken to improve the management of both terrestrial and marine protected areas, to secure permanent financing for these systems in perpetuity, and to prepare them to adapt to the challenges of climate change.

In June 2007, President Oscar Arias launched his visionary "Peace with Nature" initiative, which encompassed several environmental initiatives like the Carbon Neutrality proposal. At the invitation of the president, the Linden Trust for Conservation established a partnership with the Betty and Gordon Moore Foundation and The Nature Conservancy to work with the government on one of its initiatives: the creation of a long-term mechanism for funding the nation's protected areas. Later the Walton Family Foundation joined the effort.

A salient feature of Costa Rica Forever is that it is structured as a deal between the highest level of government and its private partners. At the more transactional level, the deal is expressed in terms of the "single closing" that took place on July 27, 2010, and launched implementation. The single closing technique has the benefits that it motivates the donors by creating leverage and a sense of urgency, ensures that all the project objectives are fully funded, and ensures that key governmental actions are taken before the funds are released. Once the project had met these milestones, a trustee, a Costa Rican NGO, was established to fulfill this role.

Conclusions

Over the last thirty years, the utilization of innovative financial instruments for conservation has been critical to complement strictly governmental efforts. The commercial and later bilateral debt-swaps make Costa Rica one of the countries that has utilized this novel mechanism more intensely to increase the support of conservation and PES efforts. For example, Costa Rica is the country that has benefited the most from the Tropical Forestry Conservation Act, having executed two debt-swaps under the agreement.

In fact, debt-for-nature swaps have been a salient characteristic of Costa Rica's conservation efforts and have led to the creation of a variety of trust funds, both public and private with a variety of governance mechanisms. In order to accomplish this, Costa Rica has established many different partnerships, with donor governments like Canada, the Netherlands, Spain, Sweden, and the United States. The partnerships have included all major international NGOs (Conservation International, TNC, WWF, etc.), private foundations, and individuals who have often supported Costa Rica in developing innovative initiatives.

Another key pillar has been the ability to establish partnerships with scientists and conservationists throughout the world, aided by organizations like the Organization for Tropical Studies (OTS). OTS has taken thousands of biologists to its courses at La Selva and throughout the country for many years, and it continues to do so. In fact, most tropical biologists throughout the world have been to Costa Rica at one time or another.

Many of these have been longtime friends and some have become Costa Rican citizens, such as the US Quakers who started the Monteverde Reserve in 1954. There are many other examples.

In summary, Costa Rica has experimented with a large number of financial instruments, and it has often blended public and private funds to accomplish conservation objectives. Many of these have been highly successful and have provided the resources and stability for critical projects. No assessment of forest conservation efforts would be complete without an analysis of how these efforts have impacted rural communities. Since Costa Rican law allows for only limited development inside the national parks, the communities surrounding the parks provide many of the services in terms of food and lodging, as well as other ecotourism opportunities like bird watching, whitewater rafting, and adventure tourism. A study by the Tropical Agricultural Research and Higher Education Center (CATIE) has shown that communities surrounding national parks have at least 10 percent more income than other similar communities.

The PES system can also be considered as a fundamental effort to transfer resources to rural areas and a large part of the program was administered through cooperatives or community agricultural centers. Over 10 percent of the program has gone to indigenous communities, which hold their land communally and also apply the program based on their priorities, such as schools or health centers.

Key References

FONAFIFO (National Forestry Financing Institute). 2015. *Annual Report*. San José, Costa Rica: FONAFIFO.

Gordon and Betty Moore Foundation, Linden Trust for Conservation, the Walton Family Foundation, The Nature Conservancy. 2011. *Forever Costa Rica*.

Perez, Isaac, and Alvaro Umaña Quesada. 1996. *El Financiamiento del Desarrollo Sostenible*. Alajuela, Costa Rica: INCAE.

Porras, Ina, David N. Barton, Miriam Miranda, and Adriana Chacón-Cascante. 2013. *Learning from 20 Years of Payments for Ecosystem Services in Costa Rica*. London: International Institute for Environment and Development.

Pringle, Robert M. "Upgrading protected areas to conserve wild biodiversity." *Nature* 546, no. 765:91–99.

Umaña Quesada, Alvaro. 1990. "Costa Rica's fight for the tropics." In *Encyclopaedia Britannica Science Yearbook*, 127–45. Chicago: Encyclopaedia Brittanica, US.

———. 2016. *Point West: The Political History of Guanacaste National Park Project*. San José, Costa Rica: CLAVE.CRUSA.

CHAPTER 14

United States: Blending Finance Mechanisms for Coastal Resilience and Climate Adaptation

Katie Arkema, Rick Bennett, Alyssa Dausman, and Len Materman

Note: The findings and conclusions in this chapter are those of the authors and do not necessarily reflect the views of the US Fish and Wildlife Service.

....

Sea level and storms are expected to increase significantly by midcentury, threatening coastal infrastructure and communities worldwide. The cost, failure, and impact of traditional, hardened shorelines has fostered interest in coastal ecosystems (often termed green infrastructure*) for risk reduction, as well as the recreational, livelihood, and other cobenefits they provide. A growing body of research and on-the-ground examples demonstrate the potential for wetlands, reefs, coastal forests, and other habitats to enhance the resilience of shorelines. However, confronting global climate change and coastal hazards requires diverse funding mechanisms to take ecosystem restoration and conservation for coastal resilience to scale. Here we share case studies from three locations in the United States that center around conservation and restoration of ecosystems for risk reduction and multiple societal benefits. These cases from the East, West, and Gulf coasts cover a variety of funding mechanisms, including liability, government subsidy, and private sector investments in ecosystems. Despite differences in financing and geography, they illustrate the importance of collaboration among multiple actors to achieve a common set of ecosystem service objectives that stakeholders want from* *coastal zones. These cases can serve as examples for other jurisdictions in the United States and around the world of the lessons learned and opportunities for harnessing ecosystem restoration and conservation—funded by diverse finance mechanisms and supported by diverse entities—to achieve coastal resilience and multiple benefits.*

....

The overarching goal of natural and nature-based approaches is to enhance the resilience of coastal communities to sea-level rise and storms, while maintaining or restoring the multiple benefits of ecosystems for people, now and in the future. Natural approaches to coastal protection include conservation of existing wetlands, coastal forests, dunes, reefs, and other ecosystems that

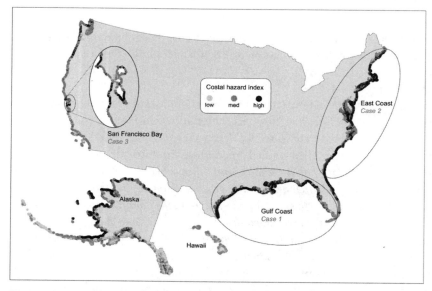

Figure 14.1. Exposure of the United States coastline and coastal population to sea-level rise in 2100 and storms. Darker colors indicate regions with more exposure to coastal hazards. Black ovals indicate the locations of the three case studies: ecosystem restoration in the US Gulf Coast, postdisaster investments for coastal resilience along the eastern United States, and watershed and floodplain planning and investments in the San Francisco Bay Area. Data depicted in the inset maps are magnified views of the nationwide analysis. A version of this figure first appeared in *Nature Climate Change* (Arkema et al. 2013).

have the ability to attenuate waves, reduce water levels, and trap sediments. Nature-based approaches include restoration of degraded ecosystems and hybrid approaches that combine natural features and traditional structures (e.g., saltmarsh coupled with rock levee). Through buffering shorelines from coastal erosion and flooding, ecosystems have the potential to mitigate the effects of coastal hazards (fig. 14.2), as well as reduce the size and cost of built defenses, such as levees behind marshes.

National, regional, and local government agencies, nongovernmental organizations (NGOs), industry representatives, and waterfront residents around the world are beginning to invest in conservation and restoration of ecosystems to achieve coastal resilience and other goals. In this chapter we describe three examples of how these investments are being made in the United States. The first example comes from the US Gulf Coast, where Hurricane Katrina

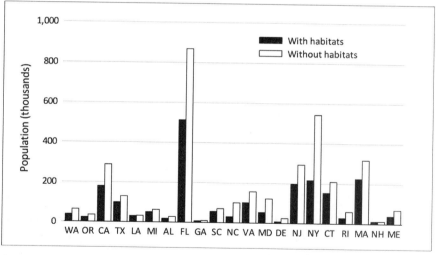

Figure 14.2. Population living in areas most exposed to hazards (the darkest shaded coastal segments in the map in fig. 14.1) with protection provided by habitats (*black bars*), and the increase in population exposed to hazards if habitats were lost owing to climate change or human impacts (*white bars*). The difference between these bars represents the number of people benefiting from reduced risk to coastal hazards provided by ecosystems. Letters on the x axis represent US state abbreviations. A version of this figure first appeared in *Nature Climate Change* (Arkema et al. 2013).

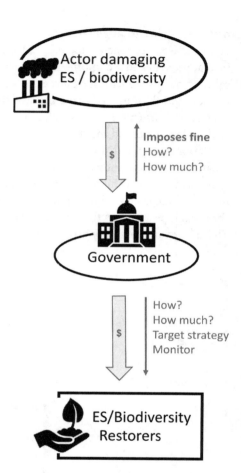

Figure 14.3. Under the liability mechanism, a government institution such as a court or agency imposes a fine on an actor damaging biodiversity or an ecosystem service. The government may receive the fine and distribute it to actors who restore biodiversity or ecosystem services, or the funds may flow directly to those undertaking restoration.

brought global attention to the role of wetlands in coastal protection, and where litigation following the Deepwater Horizon Oil Spill has provided funds for ecosystem restoration (fig. 14.3). The second example is from the northeastern United States where government subsidies following Hurricane Sandy in 2012 elevated the importance of ecosystem restoration for coastal resilience in postdisaster reconstruction (fig. 14.4). The final example is from the west coast of the United States in the San Francisco Bay Area where a single government entity working across multiple jurisdictions directs public-private

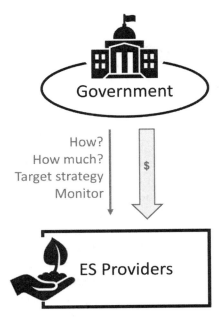

Figure 14.4. Under the government subsidy mechanism, the government compensates ecosystem service providers with funds from its general budget.

investments in ecosystem restoration to achieve coastal resilience and climate adaptation.

Case 1. Ecosystem Restoration in the US Gulf Coast

The Problem

The Gulf Coast ecosystem is vital to the people, environment, and economy of the United States, providing abundant seafood, valuable energy resources, diverse recreational activities, and a rich cultural heritage. Its waters and shorelines are home to thousands of species, including ecologically and economically important fish and invertebrates, birds, sea turtles, and marine mammals. Over 22 million Americans live on the Gulf Coast, many of whom work in crucial industries like commercial seafood and tourism. With ten of the nation's fifteen largest ports, nearly a trillion dollars in trade each year flows from the region (Gulf Coast Restoration Council 2016). Much of this value is built upon the Gulf Coastal environment and the many benefits it provides to people in the region and beyond.

These benefits, and the Gulf Coast ecosystems that underlie them, were significantly degraded by the Deepwater Horizon Oil Explosion and subsequent spill. The explosion occurred on April 20, 2010, killing eleven people working on the rig and injuring many more. The spill that followed affected more than 650 miles of Gulf Coast habitat and reduced the abundance of key species throughout the marine food web. Impacts from oil on coastal and marine ecosystems also affected key industries, leading to the closure of more than a third of federal waters to commercial fishing and an estimated US$3.6 billion in losses to the tourism sector. The degradation from the spill, and other past and ongoing natural and human activities, represents a serious risk to the cultural, social, and economic benefits derived from Gulf Coast ecosystems.

In response to the Deepwater Horizon Oil Spill, President Obama signed the RESTORE Act into law on July 6, 2012. The act calls for a regional approach to restoring the long-term health of the valuable natural ecosystems and economy of the Gulf Coast region. The RESTORE Act dedicates 80 percent of civil and administrative penalties paid under the Clean Water Act to a Gulf Coast Restoration Trust Fund (Trust Fund) for ecosystem restoration, economic recovery, and tourism promotion in the Gulf Coast region. This effort is largely federally managed.

In addition to efforts under the RESTORE Act, there are two other major restoration efforts that direct funds flowing from litigation related to the Deepwater Horizon Oil Spill. The second effort aims to restore natural resources damaged by the spill through a Natural Resource Damage Assessment (NRDA) under the Oil Pollution Act. This effort is being coordinated among Gulf states and the federal government. The third effort is being administered by the National Fish and Wildlife Foundation (NFWF), a nongovernmental organization (NGO), using monies from the settlement of criminal charges against British Petroleum and Transocean Deepwater, Inc., put into a Gulf Environmental Benefit Fund (GEBF). The three efforts are administered in different ways by different combinations of state and federal agencies, as well as one NGO, but in all cases, they seek to direct funds from the Deepwater litigation and fine process to restore ecosystems and the benefits they provide for the economy and people of the Gulf (fig. 14.3).

The Ecosystem Services

Ecosystem restoration contributes to coastal resilience from storms and sea-level rise in two main ways. First, restoring natural processes and degraded shoreline habitats can reduce coastal erosion and flooding. Ecosystems such as mangroves and other types of coastal forest, saltmarshes, oyster and coral reefs, dunes, and seagrasses attenuate waves and water flow, trap sediments, and retain soils. Second, these coastal ecosystems provide a suite of other ecosystem services, beyond coastal protection. Wetlands and reefs provide recreational enjoyment through snorkeling, kayaking, hunting, and fishing; habitats provide homes for early life-stages of key fish and shellfish, which in turn support coastal livelihoods and economies, and vegetation sequesters, stores carbon, and takes up nutrients and other contaminants, mitigating climate change and enhancing water quality.

In the case of the Gulf, if coastal protection services are not maintained through restoration and management, communities will increasingly need to rely on hard infrastructure approaches such as levees and seawalls. Such hard infrastructure can further exacerbate erosion and flooding (e.g., the levees in Hurricane Katrina) and negatively affect ecosystems that support fisheries, tourism, carbon, and water quality benefits.

Ecosystem Service Beneficiaries

The beneficiaries of reduced flooding and erosion are coastal property owners and residents, as well as people using public land along the water and the agencies that manage that public land. In addition, beneficiaries of restoring Gulf ecosystems include commercial and recreational fishing industries; the tourism sector; and recreational hunters, fishers, swimmers, and beachgoers, among others. Finally, the contractors and environmental engineering firms that implement the restoration projects, benefit from the employment opportunities.

Ecosystem Service Suppliers

Restoration investments are often on public land and in nearshore waters and include restoring coastal wetlands, beaches, dunes, oysters, and other types. However, projects also occur on private land. For example, restoration investments are made in coastal watersheds with agricultural communities or through working with more urbanized communities that are on septic systems discharging into Gulf estuaries, to get them hooked into centralized sewer systems. The Restore Council and other institutions leading restoration efforts work closely with private landowners to identify, permit, and complete restoration projects. They also work with a variety of government agencies and local entities to implement restoration on public lands and in nearshore waters, including multiple large-scale coastal wetland restoration projects throughout the Gulf.

Terms of the Exchange: Quid Pro Quo

The total funding for ecosystem and economic restoration in the Gulf Coast from the Deepwater Horizon Oil Spill amounts to more than US$20 billion, with the major funding streams being ~US$5.3 billion from civil penalties under the Clean Water Act to be dispersed according to the RESTORE Act, just over US$2.5 billion to be administered by NFWF, and no more than US$8.8 billion from NRDA. Funding has gone to a variety of restoration activities, with the NRDA and NFWF processes directly tied to restoring natural resources damaged by the oil spill, and the RESTORE Act programs focusing more broadly on connections between natural resources and the economy.

Some of the funding contributes directly to adding coastal habitat to the land- and seascape. For example, funds go to replanting of seagrass, saltmarsh, and dune grass, and construction of nearshore oyster reefs, using various techniques such as reef balls, spat, and shells. Funding also supports identifying locations suitable for habitat restoration and monitoring the biophysical changes on the landscape after restoration. In addition to direct addition of new habitat to the landscape, funds go toward activities that will indirectly restore coastal and marine ecosystems. Examples include eradication of invasive species; investment in stormwater infrastructure; protecting and restoring riparian corridors to ensure nearshore water quality; improvements necessary for seagrass and oysters; interventions to restore coastal hydrology; and projects to increase or decrease sediment, as appropriate, to address land loss or water quality and/or clarity issues. Some funds also go to economic development, job opportunities, and educational activities in coastal communities, as well as land acquisition for conservation at a fair market value from willing landowners.

Recipients of the funds are usually a variety of state and federal agencies, and in some cases, NGOs or other implementing entities based on consultation with the agencies. These government agencies are the stewards of the public lands and waters and the threatened and endangered species, and it is their job to restore the resources the public has the right to.

Mechanism for Transfer of Value

The transfer of funding to service providers depends on the funding stream. The US$5.3 billion designated under the RESTORE Act from the civil penalties under the Clean Water Act goes to the Gulf Coast Restoration Trust Fund. The restoration funding is administered by the Restore Council, an independent federal agency, as well as by the US Department of the Treasury. The Restore Council includes the governors of the five states in the Gulf of Mexico (Texas, Louisiana, Mississippi, Alabama, and Florida), and the secretaries of five federal agencies (Agriculture, Army, Commerce, Homeland Security, and Interior) and the administrator of the EPA. The RESTORE Act restoration funding managed by the Council and/or Treasury administers the funds through grants and interagency agreements, according to federal law and can only be given to the five Gulf states and/or six federal agencies on the Council. The states and agencies then implement the projects and programs, or often, in turn, pay local businesses, contractors, and environmental consulting firms to implement the work. While the federal granting process has tremendous oversight, it can also be a lengthy process to get funding out for implementation. In addition, all projects administered by federal funding are held to federal environmental laws, such as the National Environmental Policy Act (NEPA).

The funds from the criminal penalties under the Clean Water Act total approximately US$2.5 billion to NFWF and are spent on restoration related to oil spill damages. NFWF has a granting program they run and collaborate with local entities and government agencies. The process for granting is written into the criminal consent decree. NFWF is not held to federal granting law, but they have a robust review and approval process for projects. In addition, depending on the implementing entity, they are not always subject to all federal environmental laws.

Finally, the largest stream of funds over US$8 billion comes from the NRDA. In this process the states and federal government work together to decide how to invest the funds. The NRDA funding stream does not involve a granting process and thus has less oversight. However, because of the approval and involvement of federal agencies, all projects are subject to NEPA.

Monitoring and Verification

The federal government has new, extensive guidance for restoration monitoring. Every project has to submit a monitoring plan and make the data from the monitoring activities available to the public. Typically, granting programs focus on tracking restoration "outputs" such as acres acquired, built, or restored. Monitoring efforts may not capture "outcomes" from restoration, which would be defined as what the restored acres provided over the long term in terms of ecosystem function, services, and human well-being.

Effectiveness

It is too soon to know the effectiveness of the projects funded since the Deepwater Horizon Oil Spill for achieving coastal resilience and other related social and economic goals. In the past the focus has been on monitoring biophysical factors, such as area of habitat restored, but increasingly institutions and people in the Gulf are interested in the effectiveness of restoration to achieve economic and human well-being goals.

Key Lessons Learned

The Gulf Coast case study highlights several key lessons for investing in ecosystems to achieve coastal resilience. First, coordination among multiple federal agencies and states takes time, but is important to define restoration goals and find consensus with public input before moving forward to project selection. Governance among different funding entities with specified points of input from the public and NGOs is also challenging. Establishing clear governance and transparent decision making, as well as developing science-based tools to inform project selection, would expedite and improve decision making, resulting in projects with more cobenefits and public support.

Second, mechanisms to get funding out for implementation (often implemented by state and local governments contracting NGOs and private companies) need to balance adequate oversight with efficiency and effectiveness. Grants can be overly burdensome if not implemented properly. In addition, complying with grants often means monitoring "outputs," instead of "outcomes." Monitoring restoration outcomes, such as social and economic improvements, not just biophysical outputs, is important for understanding what the restoration is actually achieving beyond "acres restored" and for communicating with the public and funders.

Lastly, federal laws are created to protect the environment. But these laws can also be expensive to comply with, and therefore create inefficiencies in restoration, specifically when projects are to restore the ecosystem, not damage it. Often states and other local governments do not want federal agencies involved because it requires additional regulatory compliance. Working to streamline and find regulatory efficiencies is important for achieving multiple ecological and social objectives.

Case 2. Postdisaster Investments for Coastal Resilience along the Eastern United States

The Problem

Hurricane Sandy was one of the deadliest and costliest hurricanes in US history. When it hit the Eastern Seaboard as a Category 2 hurricane in 2012, it became the largest Atlantic hurricane on record. At least 223 people were killed, including 160 in the United States alone, and damages amounted to more than US$75 billion. A total of twenty-four states were affected, with New York, New Jersey, Pennsylvania, Delaware, Maryland, and West Virginia suffering much of the damage and loss of life. Numerous relief efforts and fundraisers emerged after the storm, including the Disaster Relief Appropriations Act passed in January 2013 (fig. 14-4).

The Disaster Relief Act included two primary streams of funding. The first was a more traditional stream of funding directed toward rebuilding based on damages caused by the storm. In contrast, the second stream was more forward looking with a goal of increasing the resilience of the Eastern Seaboard of the United States to future storms and hurricanes. This latter stream of funding was appropriated to the Department of the Interior (DOI) and a portion was subsequently managed by the National Fish and Wildlife Foundation (NFWF). Investments were targeted toward projects that would reduce communities' vulnerability to growing risks from coastal storms, sea-level rise, flooding, erosion, and associated threats through strengthening natural ecosystems that also benefit fish and wildlife.

The Ecosystem Services

As highlighted in the previous case study in the US Gulf Coast, ecosystems play a role in coastal hazard mitigation. Saltmarshes, seagrass, oyster reefs, and other shoreline habitats have the ability to reduce wave energy, attenuate water flow, and trap sediment and soil through their three-dimensional structure. By extracting wave energy and accreting sediments along the shore and underwater, they can reduce coastal flooding and erosion that would otherwise threaten coastal communities and infrastructure.

Furthermore, the same biological structure (e.g., roots and shoots of saltmarsh and seagrass, coastal dunes and beaches) that influences coastal processes also provides habitat for ecologically and economically important species. One of the mandates of the DOI, and in particular, the US Fish and Wildlife Service, is management of threatened and endangered species, interjurisdictional fisheries, and migratory birds and the habitats that support their populations. Thus, in addition to enhancing coastal resilience, a central goal of the Disaster Relief Appropriations Act was investments in restoration and conservation for ecological benefits. For example, projects involved restoration of seagrass habitat, not only for its ability to attenuate waves and stabilize shorelines, but also for its importance as blue crab (*Callinectes sapidus*) and fisheries habitat. Dams were removed to open habitat for American eels (*Anguilla rostrata*), American shad (*Alosa sapidissima*), and river herring (alewife, *Alosa pseudoharengus*), and blueback herring (*Alosa aestivalis*), as well as reduce community flooding from storm events and the potential for catastrophic dam breaches. Similarly, restoration of beaches and barrier islands that buffer shorelines from storms, provide habitat for ecologically, economically, and recreationally important shoreline creatures, such as shorebirds and the horseshoe crab (*Limulus polyphemus*). Horseshoe crabs, for example, are important to the biomedical industry. Their hemolymph is used to detect bacterial contamination of medical devices and injectable medications for people and animals; a substitute has not yet been successfully manufactured.

Ecosystem Service Beneficiaries

The goal of resilience-funded projects was to restore and strengthen coastal marshes, wetlands, and shorelines; connect and open waterways to increase fish passage and improve flood resilience; and bolster local efforts to protect communities from future storms. The primary beneficiaries of the projects implemented with funds from the Hurricane Sandy Disaster Relief Appropriations Act are coastal property owners and residents that would otherwise need to invest in hard infrastructure approaches or suffer property damages, and the broader public that uses infrastructure (e.g., schools, roads, emergency services) and open spaces within coastal areas. In addition, commercial and recreational fishermen and the associated industry benefit from more sustainable populations of key species, such as the blue crab in the Chesapeake Bay. Furthermore, a variety of people benefit from recreational opportunities fostered through restoration and conservation. Many visitors are drawn to open spaces (e.g., beaches, wetlands) and wildlife, like shorebirds in Delaware Bay, and for recreational and physical activities. Strengthening natural defenses will in turn protect homes and infrastructure, improve fishing, crabbing, birding, hiking, and other recreational opportunities, as well as support the ecotourism industry.

Ecosystem Service Suppliers

Most of the projects funded by government subsidies through the Disaster Relief Bill tend to occur on public land, especially within the National Seashores, National Wildlife Refuges, and state and local parks and open spaces. But several projects were proposed and funded on private land, especially land with obsolete and high-liability dams. In the northeast, many of the dams require significant maintenance costs, but operations cover only a small part of the costs. Thus many private landowners with dams on their property were interested in dam removal and investments in restoring habitat quality and connectivity as a way of eliminating aging infrastructure and liability for downstream safety.

Terms of the Exchange: Quid Pro Quo

The Hurricane Sandy Disaster Relief Supplemental Appropriations Act of 2013, Public Law 113-2 (also referred to as Sandy Supplemental), appropriated US$829 million, US$786.7 million post-sequestration, to the DOI to respond to and recover from Hurricane Sandy. Included in the appropriation to DOI was US$469 million, US$445 million post-sequestration, appropriated directly to the National Park Services, US Fish and Wildlife Service, and Bureau of Safety and Environmental Enforcement for response and recovering, rebuilding lands and facilities, and grants to the states for historical preservation (i.e., the traditional stream of funding). An additional US$360 million, US$341.9 million post-sequestration, was appropriated to the office of the Secretary of the DOI with authority to transfer funds to bureaus and offices and to enter into financial assistance agreements for mitigation. The act provides explicit direction to use mitigations funds to restore and rebuild national parks, national wildlife refuges, and other federal public assets with the goal of increasing the resilience and capacity of coastal habitat and infrastructure to withstand and reduce damage from storms.

Government subsidies through the mitigation stream of funding from the Relief Bill were directed at a variety of projects. Many of these involve reconnecting fragmented rivers to contribute to the recovery of species and resilience to riverine flooding. Other projects included repairing and installing culverts to allow for sufficient mixing and fish passage, and dam removal to facilitate fish passage and reduce flood risk. Some projects involved restoration of riparian habitat for critical and endangered species, invasive species removal, clearing large debris, rebuilding pools and ponds, and restoring saltmarsh and oyster reefs. For beaches and dunes, projects included restoring and enhancing beaches, stabilizing critical nesting islands, and installing living shorelines. Additional funds were granted to inform project design and coastal resilience assessment. Outcomes from these planning grants were used to inform restoration projects implemented in a second phase of the funding. Last, some of the funding was retained for administration and future cross-project monitoring.

Mechanism for Transfer of Value

Of the US$360 million appropriated to the DOI for mitigation and strategic investments in future coastal resilience, sequestration reduced 5 percent of these funds (US$341.9 million). Additionally, US$35 million were allocated to the Bureau of Ocean Energy and Management to fund sand and gravel management activities, US Geological Service for science support, Bureau of Safety and Environmental Compliance for repairs to facilities, and the Office of the Inspector General for oversight. With the remaining US$311 million, DOI provided US$207 million to support 113 DOI Bureau projects designed to reduce ecosystem and communities' vulnerability to growing risks from climate-related threats (including coastal storms, sea-level rise, flooding). In addition, DOI partnered with NFWF to administer an external competition to support similar projects led by state and local governments, universities, nonprofits, community groups, tribes, and other nonfederal entities. Through this process, US$100 million in DOI funding from the Sandy Supplemental was invested in fifty-four projects along with more than US$2.7 million in private funding leveraged by NFWF and allocated through the competitive grant program.

Pursuing projects that promote resilience and address key science and knowledge gaps will greatly improve resilience strategies, early hazard warnings, and avoid costly mistakes in restoration and mitigation actions. These efforts will better position government and the public to address challenges posed by climate change and sea-level rise.

Monitoring and Verification

Each project funded under the Hurricane Sandy Disaster Relief Appropriations Act included a plan for monitoring and verification. However, these monitoring plans were largely designed before the projects were funded and before the DOI and other agencies collaborated to develop a set of standard metrics of monitoring coastal resilience and other ecosystem service goals. Assessing the results of Hurricane Sandy resilience projects funded through DOI is critical for developing best practices, determining gaps in knowledge, sustaining or enhancing improvements in coastal resilience created by project activity, and communicating the effective use of tax dollars to the people of the United States.

To that end, DOI created a metrics expert group of physical and ecological scientists and socioeconomic experts to recommend Hurricane Sandy ecological resilience performance metrics. In order to implement the project evaluation metrics, the DOI took a novel approach and dedicated funds for later allocation to support implementation of standard metrics across projects. In 2018, the agencies instated a competitive process for US$16 million for the previously awarded projects to apply for funds to monitor and evaluate the existing projects through 2023. Results from the ecological and socioeconomic monitoring will allow DOI to evaluate already funded projects and to inform future investments in natural and green infrastructure approaches at local, state, and national levels. Ultimately, the agencies want to assess whether the overarching goals of the government subsidies through the Disaster Bill are being met—to improve ecosystem and community resilience and to factor resilience into future community planning and project management.

Effectiveness

Many of the projects funded under the Disaster Relief Appropriations Act of 2013 are just coming to completion and the DOI is in the early stages of assessing effectiveness. Restoring the resilience of the Eastern Seaboard of the United States is a massive undertaking, and developing accurate and sensitive performance metrics is a major challenge. Nevertheless, preliminary data suggest positive results. For example, reports suggest that after dam removal, various species of fish are being found upstream of the restoration project sites. Many agency scientists and program managers assumed the coastal resilience outcomes would take longer to emerge than the ecological ones. However, data collected from one of the larger saltmarsh and barrier beach restoration projects (i.e., the Prime Hook National Wildlife Refuge, DE), in which more than 1,600 hectares, 4 kilometers of beach, and 35 kilometers of channels were restored, demonstrate that piping plovers (*Charadrius melodus*), are returning, and various tern species are occupying areas from which they had been absent since the 1960s. Moreover, residents of the adjacent bay shore agricultural communities have directly benefited from the reduction of flooding from nor'easters in the past two years, and the restoration project has not experienced appreciable damage. The efforts help protect local residents from the next big storm while creating jobs, engaging youth and veterans, and restoring habitat for wildlife.

Key Lessons Learned

The Hurricane Sandy case study highlights several key lessons for investing in ecosystems to achieve coastal resilience and climate adaptation. First, communication and collaboration among jurisdictions and agencies is key to a successful restoration funding program, project implementation, and realization of shared goals. To foster collaboration, the DOI founded an executive council (including the secretaries and directors of various agencies). This executive council made the majority of programmatic decisions. In addition, a regional leadership team was established, including regional directors from the US Geological Service, US Fish and Wildlife Service, National Park Service, Bureau of Ocean Energy Management, and a technical team. Coordination is also essential across disciplines, that is, ecological, engineering, and on-the-ground practitioners, to discuss (1) what are the lessons learned; (2) what information is needed to develop design and build criteria; and (3) what are ways to address liability, risks, and other issues of concern.

A second lesson was the importance of planning for and funding monitoring and verification. To achieve the monitoring goals, a portion of the Hurricane Sandy budget was retained to support assessment and monitoring at a later date. To understand how investments in ecosystem restoration made at the project scale influence social and economic outcomes at the landscape scale, the Hurricane Sandy program developed social and economic metrics that link back to biophysical changes and metrics. The program is collecting empirical information at the project scale and using models to scale to the whole northeastern and mid-Atlantic coastline to understand the overall change in habitat for key species and changes in resilience of coastal communities as a result of restoration investments.

Case 3. Coastal Watershed and Floodplain Planning and Investments in the San Francisco Bay Area, USA

The Problem
The watershed and floodplain of San Francisquito Creek, California, encompasses approximately 130 square kilometers from the Santa Cruz Mountains to San Francisco Bay. The Bay is central to the regions' financial and tourism sectors, as well as major ports and heavy industry. San Francisquito Creek lies in the heart of Silicon Valley near the southern end of San Francisco Bay, and its floodplain is largely comingled with the Bay floodplain. Within or adjacent to these floodplains lies major regional assets related to transportation, water supply and treatment, electrical and natural gas transmission, and businesses, including the headquarters of Google, Facebook, Hewlett-Packard, and other technology giants. Surrounded by residential and commercial development, an extensive system of coastal wetlands, and other open spaces, the area around the Creek supports numerous species of animals and plants and a variety of recreational activities enjoyed by local residents and visitors. However, despite the region's natural and financial wealth, these communities have frequently experienced riverine and coastal flooding. In fact, within the poorest areas, homes sit below sea level and roofs lie below nonengineered berms that serve as levees. Because it forms the boundary between cities and counties, and because of its history of disasters, San Francisquito Creek was historically viewed as a liability. Without a single organization to manage the Creek and the surrounding watershed, and overlapping fluvial and coastal floodplains, multiple jurisdictions struggled to achieve shared objectives.

In 1998, after the largest of several twentieth-century flooding events damaged approximately 1,700 properties, five local agencies from two counties joined together. The cities of Palo Alto, Menlo Park, and East Palo Alto, the County of San Mateo, and the countywide water agency in Santa Clara County, the Santa Clara Valley Water District, created a new regional government agency. They named it the San Francisquito Creek Joint Powers Authority (SFCJPA), after the natural feature that divides and unifies them. Elected officials represent these jurisdictions on the SFCJPA Board. The Authority also employs an executive director and three professional staff, who are assisted by consultants and staff from its founding agencies. The overarching goal of the SFCJPA is to transform San Francisquito Creek and the 18-kilometer Bay shoreline of the three cities from liabilities into unifying assets by addressing the cities' shared flooding, environmental, and recreational interests.

The SFCJPA has three large projects that exemplify its emphasis on multi-benefits (flood protection, ecosystem restoration, and recreation) across multiple jurisdictions in Silicon Valley. These include the "S.F. Bay to Highway 101" and "Upstream of Highway 101" creek projects, and the "Strategy to Advance Flood protection, Ecosystems and Recreation along the Bay" project, known as SAFER Bay. These projects are financed through a combination of public and private funding, and they incorporate both natural and hardened infrastructure to enhance resilience, habitat, recreation, water quality, access to recycled water, viewsheds, and other values.

The Ecosystem Services

The SFCJPA plans, designs, and implements projects from the upper watershed to tidal marshes in a largely urbanized region. Where possible, it emphasizes natural and nature-based approaches to coastal resilience, including the creation, restoration, and conservation of saltmarsh, creek floodplain, and other ecosystems. As in the Gulf of Mexico and US Eastern Seaboard cases, ecosystems in and around the Bay can reduce the risk of coastal erosion and flooding by attenuating waves and trapping sediments. In addition, in the San Francisco Bay Area, a primary issue is accommodating flooding from rainwater, high tides, and sea-level rise. Conserved and restored open spaces can provide a place for high water to flow to keep it from flooding infrastructure and residential properties.

In addition to coastal resilience, an SFCJPA priority is to provide opportunities for recreation and connectivity between communities. The "Baylands," as the coastal wetlands and tidal creeks are fondly called, provide an opportunity for physical exercise and mental rejuvenation in an otherwise highly urbanized setting. With the headquarters of Facebook, Google, and other large employers along the shoreline, these trails provide important commuter routes that improve employee health and reduce traffic on congested roadways. The wetlands, beaches, and mudflats are also home to numerous endangered and threatened species, drawing birdwatchers and outdoor enthusiasts. In addition, the coastal vegetation can help to filter pollutants and nutrients to maintain water quality, and to store and sequester carbon for climate regulation. The importance of these cobenefits has motivated the investigation of—and it will likely enable—new approaches to shoreline protection, which is why the SFCJPA has begun to implement a variety of natural, nature-based, and hybrid approaches to reducing risk to coastal communities and infrastructure.

Ecosystem Service Beneficiaries

Beneficiaries of investments in habitat restoration and conservation include coastal property owners with infrastructure at risk of damages from flooding and erosion. Of particular importance are those people with properties in FEMA's flood hazard areas delineated in flood insurance rate maps, who are required to purchase flood insurance if they have a mortgage. Other beneficiaries include the residents of the region that commute, walk, run, birdwatch, and pursue educational activities throughout the Baylands. Many property owners adjacent to the shoreline are concerned about viewsheds and property values. Businesses in the area include Facebook's headquarters campus within a marsh, hotels, law firms, and other commercial sectors. However, not all benefits go to the private sector. In addition to the approximately 2,700 homes in the Bay floodplain, the area surrounding the Creek includes the primary highway linking San Francisco and Silicon Valley; an airport; regionally significant water supply and treatment infrastructure; gas and electrical transmission lines; as well as schools, fire and police stations, and post offices.

Ecosystem Service Suppliers

Similar to the other two case studies, the investments in conservation and restoration of coastal habitats and traditional flood control structures (e.g., levees and floodwalls) are largely on public lands. However, SFCJPA projects are required to acquire some key private holdings, including easements on several private properties to implement the S.F. Bay to Highway 101 and Upstream of Highway 101 projects, and the engagement of Facebook and other private developers along the Bay shoreline. For its S.F. Bay to Highway 101 Creek project, in an area that was marsh prior to human development, the SFCJPA has recreated marshland within a new broad channel by acquiring land from an adjacent eighteen-hole golf course (which was subsequently reconfigured into a smaller, more ecologically friendly footprint). Other land transactions required for this now complete project include a post office parking lot, private school, autobody shop, storage facility, and other holdings that, in several cases, contained hazardous materials.

Terms of the Exchange: Quid Pro Quo

For the planning, design, and construction of the projects listed above, over the past five years the SFCJPA has raised over US$100 million, including US$50 million from local cities and water districts in two counties; US$34 million from the state's water and transportation agencies; and over US$10 million from private sources, including a utility and Facebook. These sources have provided much greater funding for SFCJPA projects than the federal government, which is a traditional source of funding for major water infrastructure projects (though it remains a long-standing partner of the SFCJPA). The SFCJPA is pursuing additional funding for its Creek and SAFER Bay projects from new sources, including an innovative approach to community-wide private flood insurance, a Bay Area–wide shoreline restoration tax, a new finance district, and the carbon cap-and-trade market.

Funded activities and structures include widening of the creek into the Palo Alto Municipal Golf Course, for which US$3 million was paid to that city. Other examples are private land acquisition including permanent and temporary construction easements. In one case, in lieu of a cash payment for an easement, the SFCJPA is moving a fire hydrant off of a private property, which benefits that business, the local fire department, and nearby properties, thus illustrating the flexible approaches the SFCJPA takes toward project financing and implementation. At the Bay shoreline of the three cities in which it works, the SFCJPA has estimated the potential for almost 7 square kilometers of restoration of ecosystems for coastal resilience, which they refer to as *green infrastructure*. An example of this green infrastructure, which the agency has already built in a marsh of high ecological value and intends to build in other areas, is a "horizontal" levee. A horizontal levee has a very gradual slope into the marsh with several habitat zones corresponding to different tidal elevations. This new approach to shoreline levees can help to sustain both marsh and levee as sea-level rises, capture and store more carbon, and reduce levee height and cost. The SFCJPA's completed Creek project will protect against the maximum possible flow to reach that previously flood-prone area and a sea level 3 meters above today's average high tide in this tidally influenced reach. The agency's adjoining SAFER Bay project along the shoreline of the two counties will mirror that level of protection against sea-level rise to significantly reduce flood risk for more than 5,000 properties across 18 kilometers of shoreline.

Mechanism for Transfer of Value
In the case of the private easements, the transfer of funds is most often made through a single check to a private landowner. In only one case was a payment made for public land, to the city of Palo Alto for the use of municipal golf course land for the widened creek channel. In several cases, the SFCJPA or a private entity provided in-kind contributions to realize a gain in the long-term value of its assets or property.

Monitoring and Verification
The projects include funding for monitoring species and habitats, especially to assess compliance with the Endangered Species Act. In general, the existing monitoring plans tend to focus on ecological metrics. Tracking human well-being and economic information related to recreation, access to open space for disadvantaged communities, and coastal resilience metrics would help to elucidate the societal outcomes of coastal habitat restoration.

Effectiveness
It is still too soon to know the outcome of the SFCJPA projects from a bio-physical and social perspective. However, from a governance perspective the intergovernmental approach to raising and allocating funds across jurisdictions appears to have traction with the many public and private interests in the region.

Key Lessons Learned
The San Francisco Bay case study highlights several key lessons for investing in ecosystems to achieve coastal resilience and climate adaptation. Similar to the other cases, coordination among multiple local, state, and federal agencies takes time, but it is important to define restoration goals and find consensus

before moving forward with selection, design, and implementation of large, regionally significant projects. The direct involvement and support of elected officials at the three levels of government is key to creating and sustaining the support of governmental agencies.

Second, funding for large, multijurisdictional, multibenefit projects can be secured from the project's various beneficiaries, both public and private—this broad sharing of the costs and benefits leverages their investments. Frequently, the private sector views public works (including environmental) projects as the responsibility of the public sector; however, when private entities accrue unique and direct benefits from a project, these entities should participate in the planning, design, and funding of these projects. The same is true for permitting. Even in settings with substantial constraints related to jurisdictional boundaries, land use, sensitive species, utilities, and other infrastructure, projects can be permitted and built that achieve substantial protections against flooding and sea-level rise, and enhancements to ecosystems and recreation. Just as with financing, multiple beneficiaries of regional projects can drive a broader view that involves compromises among a greater number of stakeholders. While challenging and time consuming, this results in projects that implement a broader vision for the community.

Along much of the Bay in Silicon Valley, restoring the ecological function of the shoreline and protecting people and property are codependent. This means that shoreline marshes can and should be created and utilized for flood protection in ways that sustain the marshes in the face of sea-level rise.

Conclusions

Together, the Gulf Coast, Eastern Seaboard, and San Francisco case studies illustrate several important insights for natural capital assessment and coastal resilience planning and financing. First, these cases highlight diverse mechanisms for funding conservation and restoration of ecosystems to achieve the multiple benefits people want from coastal systems. Government subsidies, liability mechanisms, and public-private sources all hold promise for expanding funding directed at safeguarding and enhancing ecosystems for coastal resilience, beyond more traditional philanthropic sources. Regardless of the mechanism, it is important to find a balance between adequate oversight and efficiencies and effectiveness.

Second, these cases highlight the importance of coordination among multiple entities, including government agencies, the private sector, NGOs, and other stakeholders. While coordination takes time and resources, engagement is critical to defining shared goals and building consensus. Multiple beneficiaries of coastal restoration and conservation projects can come together to overcome funding, permitting, and jurisdictional challenges. The private sector is an important player in these processes, especially when private entities accrue unique and direct benefits from public projects.

Finally, these cases show it is possible to fund and implement projects that both restore ecological function and reduce risk. Whether the long-term outcomes meet these diverse goals requires monitoring both ecological attributes and social and economic results. However, early results are promising—several Hurricane Sandy projects suggest once these projects are funded and implemented, flooding from coastal storms has waned and threatened and endangered species have returned.

Key References

Abt Associates. 2015. "Developing socio-economic metrics to measure DOI Hurricane Sandy project and program outcomes." https://www.doi.gov/sites/doi.gov/files/uploads/Socio_Economic_Metrics_Final_Report_11DEC2015_0.pdf.

Arkema, Katie K., Greg Guannel, Gregory Verutes, Spencer A. Wood, Anne Guerry, Mary Ruckelshaus, Peter Kareiva, Martin Lacayo, and Jessica M. Silver. 2013. "Coastal habitats shield people and property from sea-level rise and storms." *Nature Climate Change* 3, no. 10:913.

Department of the Interior (DOI) Metrics Expert Group. 2015. *Report for the Department of the Interior Recommendations for Assessing the Effects of the DOI Hurricane Sandy Mitigation and Resilience Program on Ecological System and Infrastructure Resilience in the Northeast Coastal Region.* https://www.doi.gov/sites/doi.gov/files/migrated/news/upload/Hurricane-Sandy-project-metrics-report.pdf.

Disaster Relief Appropriations Act. 2013. Public Law 113-2. https://www.congress.gov/113/plaws/publ2/PLAW-113publ2.pdf.

Gulf Coast Ecosystem Restoration Council. 2016. "Comprehensive Plan Update 2016: Restoring the Gulf Coast's Ecosystem and Economy." https://www.restorethegulf.gov/comprehensive-plan.

Hurricane Sandy Coastal Resiliency Competitive Grant Program. http://www.nfwf.org/hurricanesandy.

National Fish and Wildlife Foundation. 2015. "Developing socio-economic metrics to measure Hurricane Sandy project and program outcomes." https://www.doi.gov/sites/doi.gov/files/uploads/Socio_Economic_Metrics_Final_Report_11DEC2015_0.pdf.

Resources and Ecosystems Sustainability, Tourist Opportunities, and Revived Economies of the Gulf Coast States Act (RESTORE) Act. 2012. Subtitle F of Public Law 112-141. https://www.treasury.gov/services/restore-act/Documents/Final-Restore-Act.pdf.

San Francisquito Creek Joint Powers Authority. sfcjpa.org.

Sutton-Grier, Ariana E., Rachel K. Gittman, Katie K. Arkema, Richard O. Bennett, Jeff Benoit, Seth Blitch, Kelly A. Burks-Copes et al. 2018. "Investing in natural and nature-based infrastructure: Building better along our coasts." *Sustainability* 10, no. 2:523.

CHAPTER 15

United Kingdom: Paying for Ecosystem Services in the Public and Private Sectors

Ian Bateman, Amy Binner, Brett Day, Carlo Fezzi, Alex Rusby, Greg Smith, and Ruth Welters

Natural capital delivers a wide range of ecosystem services, the majority of which are "public goods," in that no one can be excluded from enjoying their benefits, and use by one individual does not reduce availability to others. While these characteristics can make such public goods of great value, they also mean that private companies find it difficult or indeed impossible to make money from such goods, and as they are often costly to produce (either directly or because it means other profitable activities have to be forgone) they are commonly underprovided. Here we show how the introduction of payments for ecosystem services (PES) can incentivize private businesses to provide public goods. We present three case studies from the United Kingdom illustrating the flexibility of PES schemes. The first two of these provide, in turn, a national level and then catchment level application of the more common form of PES scheme where private providers are funded by the public sector. The third and final case study again operates at the catchment level but now presents a more unusual variant of PES scheme funded by the private sector in a situation where the production of public benefits is a (welcome) byproduct of the production of private benefits for the funder. Together these form a matrix of funding-source and decision-level exemplars that provide wide applicability to a variety of contexts.
....

The research literature is replete with analyses showing how changes to natural capital and hence the supply of ecosystem services can enhance both environmental sustainability and public benefits. However, many of the key re-sources necessary to deliver such improvements are in private ownership. This causes a problem when, as is often the case, a move to supply greater ecosystem services incurs costs (including forgone profits) for the private resource owner. So, for example, reducing farm pesticide use may enhance river and drinking water quality but incurs costs for farmers who now have to suffer greater pest damage to the crops they rely upon for an income. Overall, society might ben-efit substantially from such a move but the farmer would bear almost all of the costs and is therefore understandably resistant. To overcome such problems a variety of payment for ecosystem services (PES) schemes have been developed to stimulate the efficient delivery of key, high-value ecosystem services that are either not produced, or are underprovided by the normal market activities of private producers.

This chapter considers two PES mechanisms: payments from the public sec-tor to private businesses, and payments between private businesses. Public-to-private funding provides the most common PES mechanism in most other countries. By contrast, private-to-private (i.e., business-to-business) PES mechanisms remain relatively novel; yet, because they tap into private sec-tor funds, they have great potential for incentivizing environmental improve-ments, particularly in cases where there is a profit opportunity arising from such improvements.

We illustrate permutations of these mechanisms through three case stud-ies: (1) public-to-private funding of natural capital improvements for national-level decision making; (2) public-to-private funding of natural capital im-provements at catchment level; and (3) business-to-business funding of natu-ral capital improvements, again at a catchment level. All three examples are taken from the UK, which has recently launched a 25 Year Environment Plan with the overarching objective to ensure that this is the first generation to leave the natural environment in a better state than it inherited (H. M. Government 2018). However, the approaches to PES presented here have relevance to coun-tries around the world.

Case 1. Public-to-Private Funding of Natural Capital Improvements: National-Level Decision Making

Following in the wake of the landmark Millennium Ecosystem Assessment (2005), the UK determined to undertake a National Ecosystem Assessment

(UK–NEA 2011)—the first comprehensive assessment of the country's eco-systems and a systematic environmental and economic analysis of the benefits they generate. The UK–NEA sought to assess the consequences of natural capi-tal use and land-use change and showed that over 30 percent of the ecosystem services provided by the UK's natural environment are in decline.

The UK–NEA data provided highly detailed, spatially referenced, time series, environmental data covering all of Great Britain, ranging from soil characteristics (e.g., susceptibility to water logging); climate variables (e.g., temperature and rainfall); and land use (e.g., agricultural output). This was complemented by similar spatially and temporally referenced data on market variables (e.g., prices and costs) and policy (e.g., incentives such as subsidies, and regulations such as land-use constraints). The resultant analysis linked environmental, policy, market, and other economic factors to examine both the market and nonmarket consequences and values generated by land-use changes. The spatially sensitive nature of these analyses also demonstrated how future policy can be targeted to most efficiently allocate available resources to maximize their net benefits.

A key element of this effort was to provide decision makers with an under-standing of the factors that drive change in natural capital resource use. A com-mon failing is that analyses consider the advantage of moving from current to alternative resource use without consideration of how the move between these two states is to be effected. The UK–NEA examined alternative land-use futures to 2060. It highlights the advantages of valuing both the market and nonmarket effects of change and the importance of spatially targeted policies. This example also illustrates the importance of understanding the environ-mental, economic, and policy drivers of change (Bateman et al. 2013), thereby allowing decision makers to see where a shift in a particular policy is likely to result in changes in land use, ecosystem services, and values.

The Ecosystem Services

The analyses assessed a mixture of market and nonmarket ecosystem services:

- Food output provides the key, market-valued, ecosystem service deter-mining approximately 75 percent of land use in the UK, including crop-land, grassland, mountain, moor, and heathland environments.
- The sequestration of greenhouse gases (GHG) has a nonmarket value and includes the direct and indirect emissions from land use and land

management, annual flows of carbon from soils due to land-use changes, accumulations of carbon in terrestrial vegetative biomass, and emissions.

- Open-access recreational visits have a nonmarket value that varies across environments (e.g., mountains, coasts, forests, urban green spaces, city centers, etc.) and locations (with visitation declining with increasing distance from populations).
- Urban green space has a nonmarket value reflecting aesthetic, physical, and mental health; neighborhood, noise regulation, and air pollution reduction benefits.
- Wild bird species diversity was used to represent biodiversity across the UK because birds are high in the food chain and are often considered to be good indicators of wider ecosystem health.

Ecosystem Service Beneficiaries
The same change can yield very differing consequences to different groups.

- Farmers: While climate change will deliver a negative impact on agricultural production in many areas of the world, the generally cool climate of the UK means that an increase in temperatures is likely to increase agricultural production and profits in some areas. However, this is likely, in turn, to negatively impact water quality and hence water supply customers, as higher agricultural intensification leads to increased nutrient pollution in waterways necessitating higher treatment costs. Lower river water quality will also impact freshwater biodiversity negatively.
- Recreationalists: Open-access recreational sites have a benefit to individuals who visit them.
- Urban residents: Typically increasing access to urban greenspace generates significant social benefits. However, the distribution of benefits can be uneven and can result in the gentrification of areas, which has the potential to push poorer families out to less-advantaged areas.
- Biodiversity beneficiaries: Improvements in wild-species diversity benefit not only the species directly or indirectly conserved (e.g., through food chains), but also any people who value such improvements. These include people who hold use values (e.g., those who engage in hunting or fishing as well as those who enjoy viewing wild species) and those who hold nonuse existence values. Biodiversity also indirectly delivers value through roles such as pollination and through diverse contributions to the maintenance of ecosystems.

- National and global beneficiaries: Both the conservation of biodiversity and the reduction of climate change through the sequestration of greenhouse gases through land-use change not only benefits the UK but has spillover effects worldwide.

Ecosystem Service Suppliers

Ecosystem services can be supplied by either the public or private sector. In the UK, private agricultural landowners and farmers are key suppliers because they control the large majority of land.

Terms of the Exchange: Quid Pro Quo

Modeling of six different policy scenarios, from 2010 to 2060, explored the effects of each on land use across Great Britain. For simplicity, here we consider the two most extreme policy scenarios. The World Markets (WM) scenario prioritizes economic growth by completely liberalizing trade. Here trade barriers are dissolved, agricultural subsidies disappear, and as a result farming moves toward large-scale, intensive production methods. By contrast, the Nature@Work (NW) scenario's main priorities are enhancing and maintaining the output of all ecosystem services and adapting to climate change.

Ignoring all other values, an analysis of the market priced outputs of land suggests the WM scenario is optimal as this maximizes intensive agricultural use of the land. However, once nonmarket values are considered, the NW scenario is clearly shown to be better for social value while the WM world actually reduces overall value and is particularly adverse for biodiversity. In both cases the spatial targeting of land-use decisions greatly increases their efficiency and, in particular, raises the "value for money" (i.e., the level of benefits relative to costs) of public spending.

Mechanism for Transfer of Value

The most common method for transferring farming-related values in the UK is via government-funded subsidies and indeed these currently make up over 50 percent of UK farmers' income. At present most of this funding is paid on a per hectare basis with little link to the ecosystem services generated. However, the decision to leave the EU and hence its Common Agricultural Policy (CAP) has led the UK government to announce that it will adopt a new "public money for public goods" approach to agricultural support (Defra 2018). This has the potential to very significantly improve the value for money generated under PES schemes, particularly if they are targeted toward those areas that yield the highest benefits (Bateman and Balmford 2018). By mapping which areas offer superior value for money for agriculture, and which would be best turned over to the production of other ecosystem services, and targeting funding accordingly, the overall efficiency of subsidies can be greatly enhanced. From a policy perspective, the adoption of such spatial targeting offers a major and potentially costless increase in taxpayer value for money simply by redirecting the current level of subsidy according to the benefits it generates.

Monitoring and Verification

PES under both current and potential UK agricultural policy are from the public-to-private sector (here, farmers). One important PES, the UK Countryside Stewardship Scheme (CSS), opened for the first time in 1991 covering only 25,000 hectares. By the time it was paused in 2004, the scheme covered over 530,000 hectares, and its payments to farmers were in excess of £52 million per year. The scheme was aimed at sustaining the beauty and diversity of the rural landscape and offered incentive payments to farmers for the preservation and provision of wildlife habitat. It made payments to farmers for arable land reversion, establishing and maintaining grasslands, as well as managing and preserving footpaths and bridleways. Participating farmers had to sign a contract, which usually lasted ten years and required them to maintain certain land uses for that time. They were monitored through a series of checks including farm visits to see if the farmers were in breach of their contracts, and if they were not providing the ecosystem service, their contract was terminated and fines were sometimes issued. The payment authority set an objective to inspect 10 percent of the farms every year, but the actual inspection rate never reached this.

Further improvements could be achieved through a move toward payments by results. In some cases, this may be reasonably straightforward to implement; for example, the quality of water running off a field is something that, in some respects (e.g., pesticide or fertilizer content) a farmer can influence. However, other payment-by-results schemes are more difficult to implement. For example, with respect to wildlife, a farmer may provide food sources for birds to thrive, yet the bird roosts on another farmer's land. Questions then arise over who should be paid for such services. Despite this, payment by results has important advantages over the input-based approach:

- Outcome-based approaches incentivize land use that will produce the best environmental result.

- It allows farmers more flexibility in the management of their land because restrictions imposed by the input-based system are removed. Appropriately designed, this could lead to a greater uptake of participants in the scheme and greater improvements if targeted appropriately on expected environmental outputs.
- Farmers are permitted to innovate and incorporate their existing knowledge into environmental provision, which can lead to a greater efficiency of production. Extensions of this knowledge could be delivered through training schemes.

The ongoing incorporation of drone-based information, earth observation, satellite data, and other "big data" systems to monitor farmland uses and certain consequences (e.g., tree planting and growth) offer further considerable potential for enhancing PES schemes.

Effectiveness

Agri-environment PES schemes have been proved to be successful where they are well focused upon ecosystem service provision such as the establishment of habitat for breeding birds (Baker et al. 2012). However, such targeting is the exception rather than the rule under a system where the majority of subsidies are allocated on a per hectare basis. This has to be balanced against the difficulties of administering targeted schemes such as CSS, which has been the subject of considerable criticism in terms of the high administrative burdens they place upon farmers.

Key Lessons Learned

This analysis highlights the following key considerations when making decisions about land use:

- Analysis of the policy, market, and environment drivers of change allows decision makers to understand which levers of change are available and which drivers are exogenously determined.
- As much as possible, capture all of the major effects that changes in land use can have upon ecosystem services to avoid misleading conclusions.
- Ensure that the different units used to measure changes in ecosystem services are commensurable, often most readily through monetary values.
- Account for values that cannot be robustly assessed in monetary terms.

Biodiversity is a particular challenge, and we argue that generally agreed-upon objectives (drawn in part from legislation but also from other preference measures and informed by ecological knowledge) should be incorporated into analyses by identifying and rejecting investments that violate those objectives. The costs of securing these objectives can be estimated and the effectiveness of options for their delivery assessed, but such costs should not be interpreted as indicators of the value of delivering sustainable biodiversity or other objectives.

- Account for spatial variation by examining alternative targeting for action and investments.
- Account for temporal variation by ensuring that changes in drivers (for example, those arising from the effects of climate change) are incorporated into analyses.
- Explore alternative policy scenarios to see the varied outcomes that can arise from different policies.

Case 2. Public-to-Private Funding of Natural Capital Improvements: Catchment-level Decision Making

While national-level decision making is vital for directing overall and longer-term strategy, most practical environmental management decisions are made at a more restricted local scale. Nevertheless, the need for environmental coherence in decisions remains, and in recognition of this there has been a noticeable movement toward consideration of catchments as an important decision-making unit. Irrespective of the scale, however, the key principle for decision making is that all of the major effects of a proposed change should be incorporated into analyses, including relevant spillover benefits and costs arising outside the immediate locus of activity.

This case considers improvements to conventional public-to-private PES decisions at the catchment level by analyzing the effects that climate change is likely to have upon the Aire catchment in Yorkshire, UK. The increase in average temperature generated by climate change induces land-use responses as farmers exploit the potential for increased agricultural intensification. This, in turn, generates higher food production but needs to be balanced against the associated impacts of increased nutrient application, higher river pollution, and lower ecological quality and recreational values. Full details of this case study are provided in Bateman et al. (2016).

The Ecosystem Services

The 86,000 ha catchment of the River Aire encompasses highly heterogeneous land uses, water qualities, and socioeconomic characteristics. The upstream western half of the catchment has low population density with its mainly upland landscape being dominated by rough grazing and pastoral agriculture. The downstream eastern half of the catchment has some mixed and arable farmland but is dominated by high-density urban areas that are themselves a major determinant of river ecology and the main source of recreational visitors.

With rural land use ranging from extensive pastoral to intensive arable farming across the catchment, the Aire is subject to significant levels of diffuse pollution including high nutrient loadings. The arable and root crops produced in the catchment are strongly and positively associated with high levels of eutrophication, which also rises in areas of high-stocking density. Conversely, extensive grassland systems are associated with less eutrophied waterways.

Ecosystem Service Beneficiaries

The general public benefits from the river in two ways. First, they have a use value through river-related recreation. This includes walking, running, and cycling along the riverbank, nature watching, canoeing, fishing, and swimming. Second, they have a nonuse value associated with the biodiversity levels in the river catchment, quite separate from their direct enjoyment of wild species. Water quality improvements also benefit water companies through reductions in treatment costs.

Ecosystem Service Suppliers

Farmers and landowners in the river catchment exert a major influence over river water quality. Climate change is predicted to induce increased intensification of agricultural production and hence greater applications of inputs such as fertilizer. This will in turn reduce river water quality through diffuse pollution. Conversely, farmers could maintain or even improve water quality by adopting more extensive production methods.

As such, farmers are the potential suppliers of improved water quality. This in turn would avoid potential losses of recreational value and other associated ecosystem services. However, this would incur income reductions for those farmers. A key objective of this study was to calculate the forgone income of, and hence compensation required by, farmers who avoid a move toward higher input, intensive agriculture. This cost is then compared to the value of recreation and other benefits maintained by such action.

Terms of the Exchange: Quid Pro Quo

We start off by considering the drivers of change. We hold market drivers constant so prices track general inflation levels and remain constant in real terms. We then simulate change in the environment driver, specifically by assuming a 1°C increase in temperature across the region and some shifts in the seasonal pattern of rainfall. Subsequently, we also consider the effectiveness of policy drivers designed to counteract the negative consequences of climate change and improve water quality.

As noted in case 1, while global climate change will place greater stress on the world agricultural system, within certain temperate countries such as the United Kingdom, increases in temperature are expected to result in elevated potential for agricultural production. Our analysis suggested that farmers will shift away from livestock production and toward cropping regimes that are expected to become more profitable, resulting in an aggregate increase in farm income of roughly £11 million per year (calculations in Bateman et al. 2018). However, this shift in production involves a greater use of some farm inputs, notably fertilizers. This in turn would increase diffuse pollution into waterways and rivers, elevating their nutrient levels and eutrophication and causing a decline in the ecological quality of the river. This not only reduces associated nonuse values, it also generates a potentially major nonmarket externality in terms of the impact it has upon the recreational value of visiting the river. By applying spatially sensitive methods for valuing such externalities we generate estimates of the geographical distribution of these losses as illustrated in the upper row of figure 15.1. Aggregating across the study area suggests that the diffuse pollution generated by the shift toward more intensive agricultural production causes a loss of some £26 million per year in recreational value. Clearly a PES that fully compensates farmers for foregone profit would generate a net benefit of some £15 million per year.

The lower row of figure 15.1 illustrates the geographical distribution of a further policy intervention, which not only avoids the losses associated with greater intensification but also increases water quality from its current low standards around the urban hubs of Leeds and Bradford up to pristine levels. A policy to both avoid a climate change–induced reduction in water quality from present day levels and then improve that quality to pristine, would yield a net benefit of £74 million per year (Bateman et al. 2018), a figure that reflects the high populations and hence recreational demand of those cities as reflected in the maps of the lower row.

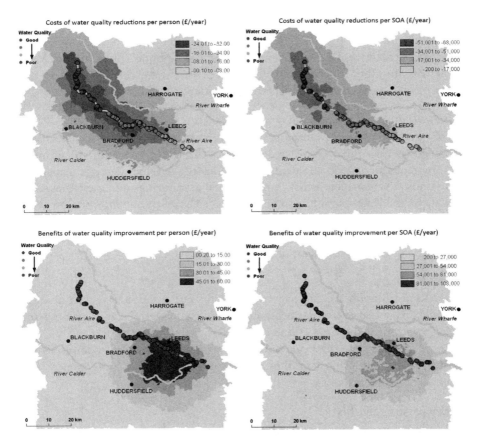

Figure 15.1. The spatial distribution of per-person (*upper left-hand panel*) and per-Census Super Output Areas (SOA) aggregate (*upper right-hand panel*) costs of water quality reduction as a result of climate change. The spatial distribution of per-person (*lower left-hand panel*) and per-SOA aggregate (*lower right-hand panel*) benefits of water quality improvement as a result of policy intervention.

Note that dots represent water quality at sites along the River Aire varying between pristine, 'good and poor quality. For a colored version of this figure, please see the version of this paper given at http://www.exeter.ac.uk/leep/publications/workingpapers/. In the *upper row*, dots show the water quality at present; note that sites currently with pristine quality will decline to good quality as a result of climate change. In the *lower row*, dots show water quality following implementation of an improvement policy where all sites attain pristine quality. Note that both the current water quality (*upper row*) and that expected as a result of climate change are predicted from the model of chlorophyll-a reported in Bateman et al. (2016), whereas water quality under the policy response scenario (*lower row*) is assumed to attain the objectives set out under the EU Water Framework Directive. (*Upper row* from Bateman et al. 2016; *lower row* from new calculations).

Mechanism for Transfer of Value
Funding for the improvements outlined above should be taken from more targeted use of current agri-environmental or water stewardship schemes. Alternatively, funding could be obtained from private water companies through a mixture of regulation and incentives (see case 3).

Monitoring and Verification
The general mode of monitoring and verification employed in the United Kingdom for similar land-use decisions is through a mix of in-stream measurement and occasional on-site inspection. The development of cheaper water monitoring mechanisms—such as digital monitoring equipment that relays data in real time—should also be encouraged to distribute information about water quality to both farmers and monitoring authorities. Such information could enable preventative measures, such as closing water abstraction inlets with prior warning of major pesticide incidents.

Effectiveness
We used a simplified climate scenario for illustrative purposes. Future analyses could make use of a wider array of climate change dimensions and dynamics, including nonlinear effects and uncertainties. Effectiveness analyses are complicated by the timescale of environmental changes. Lags commonly occur between policy decisions, implementation, behavioral consequences, and environmental response.

Key Lessons Learned
While acknowledging that the direct market impacts of climate change on agriculture may be positive within the UK context (although not in many other countries), a central contribution of this research has been to demonstrate that focusing solely on these direct impacts paints a highly incomplete picture of the net impact of climate change. Several key points emerge from examining the case of the River Aire:

- It is important to understand and model both the direct and indirect effects that climate change may have on a catchment area. Within the River Aire catchment these include the following :
  - o Direct impacts on the ecological quality of a river, induced by a temperature rise of 1°C, increasing water temperature, and in turn negatively impacting the native flora and fauna of the river.

○ Indirect impacts on the ecological quality of a river through induced land-use change, causing a shift toward crops, higher pollution of waterways, and an associated decline in recreational values of the river.
- Policies to offset the negative indirect effects of agricultural shifts can be designed so that benefits significantly outweighed costs.
- Methods for funding such change include both direct state support and business-to-business arrangements.

Case 3. Business-to-Business Funding of Natural Capital Improvements: Catchment-Level Reverse Auctions

Water quality in rivers is of significant interest to private water companies that spend large amounts of money treating water to conform to legal standards. Stopping pollution at the source by paying farmers to reduce their inputs of pesticides, fertilizers, and particulates can dramatically reduce this cost. There are also public benefits to reducing pollution as it results in increased levels of biodiversity, recreational and tourism benefits, fish catch, and reduced incidence of siltation and associated costs. The challenge is to design and implement policies that realize these win-win outcomes.

The Ecosystem Services

Figure 15.2 shows the Fowey River Catchment in South West England that the local water company, South West Water (SWW), uses to supply 95 percent of Cornwall's drinking water. However the upstream abstraction area is also used for a variety of agricultural production, which results in substantial levels of diffuse pollution and siltation into the river. The area is also important for tourism and local recreation.

Ecosystem Service Beneficiaries

At present, farmers are the main beneficiaries of the high levels of agricultural inputs that are the source of diffuse pollution in the Fowey. In contrast the water company and its customers bear the costs of treating water while wider society also suffers the externalities associated with water pollution. Compensating farmers for investing in low-pollution technologies and capital would at worst need to leave them no worse off. The resulting reduction in pollution would benefit the water company and its customers in terms of lower treatment costs and hence reduced water bills. Moreover, there are positive spillover effects to

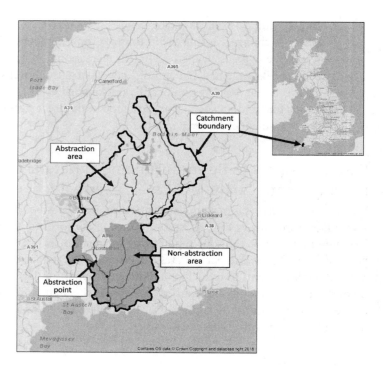

Figure 15.2. The Fowey Catchment, Cornwall, UK.

the environment and society. The environment benefits from the natural capital market to reduce the levels of diffuse pollution in the river. Lower chemical loading in the river increases levels of dissolved oxygen and reduces eutrophication. Consequently, native flora and fauna, such as waterlilies, thrive, giving greater diversity to the ecosystem. Society also benefits from such improvements through enhanced recreation, greater economic activity (e.g., through greater viability for commercial fisheries and shell fisheries, tourism growth, etc.), reduced costs (e.g., regarding the costs of dredging ports and waterways), and improved nonuse values (e.g., from improvements in wild species habitats and associated biodiversity).

Ecosystem Service Suppliers
Any landowners who live near the river are potential suppliers of clean water to SWW. The majority of these landowners are farmers who are also the main polluters of the river.

Terms of the Exchange: Quid Pro Quo

SWW pioneered the Fowey River Improvement Scheme as part of their Up-stream Thinking Initiative, a program of pollution prevention rather than remediation. Working with various of the authors (Day et al. 2013), key deter-minants of any intervention included the following: (1) What the change was: the capital investments available including slurry storage, excluding livestock from watercourses, covering livestock feeding areas, and so forth; (2) Impacts on water quality: the effectiveness of each investment. A major element of this analysis concerns the location of any change, proximity of water courses, and pollution transport determinants such as soil characteristics and gradient; and (3) The likely costs of change, including the amount of capital investment re-quired to implement the change by both the farmer and SWW, and the cost of monitoring that the change has taken place.

Mechanism for Transfer of Value

A key determinant of the transfer of value was a prior decision by the UK water regulator, Ofwat, to allow water companies to engage directly with farmers to prevent the pollution incidents arising rather than purely treating water that had already been polluted. With this in place, Day et al. (2013) designed a competitive, reverse-auction PES market that worked as follows:

- Water quality researchers provided information on the transport of pol-lution from all areas of the catchment, identifying the impacts of poten-tial changes across locations.
- Farmers were then invited to prepare bids for investments from the wa-ter company for improvements to water-related infrastructure on their farms (e.g., better drainage and containment of animal effluent). All farmers in the catchment were approached, thereby setting up competi-tion for investments among farmers and driving down stated costs.
- All bids were compared with their expected benefits to allow the water company to derive a value-for-money score for each potential invest-ment. Those bids delivering the highest value for money (i.e., lowest cost per unit of benefit) were funded.

To further drive down costs, the market allowed "practice rounds" of bid-ding. Here, once bids were received, farmers were informed where their bids would not have attained the necessary value for funding and allowed to revise their bids, either by reducing costs or altering investments to improve benefits.

After two such practice rounds, a final "for real" round was undertaken, contracts were awarded, and investments made.

By inviting multiple farmers to submit bids, the buyer sets up a competitive market where farmers now have an incentive not to overstate their compensation requirements. Cash payments were made directly from SWW to funded farmers so that the capital investments could proceed promptly. These payments were made on the condition that the farmers signed and fulfilled a contract, typically for twenty-five years for more substantial, longer-lived investments (e.g., slurry storage and roofing) and ten years for shorter-lived capital items (e.g., fencing and concreting). The contracts stipulated that the capital items in question would have to be used for the agreed purpose, be properly maintained, and insured against any damage.

Monitoring and Verification
All of the capital investments were easy to monitor and verify because payments were issued on the actions taken by farmers. Visible confirmation of a build or management changes was all that was needed.

Effectiveness
The reverse-auction case study discussed here was a relatively low-budget, proof-of-concept exercise that proved very successful and provides positive prospects for similar schemes to be run in the future. SWW had £360,000 of funds to distribute but was considerably oversubscribed, receiving bids for £776,000 worth of investment. The reverse auction approach to investment decisions was also 20–40 percent more cost-effective than a prior noncompetitive scheme where advisors allocated funding on the basis of farmers' costing statements. It is expected that over time these cost reductions might erode somewhat as farmers gain more information on the operation of the PES market. However, the administrative ease of the market approach combined with residual competitive efficiency makes it an appealing alternative.

The limitations of this scheme centered around the difficulty of assessing the impacts the capital investments had on water quality. First, it was hard to know whether the investments made were ones that would deliver the best water quality improvements on the farm. All of the investments were reviewed by an advisor and deemed to be ones that would deliver environmental improvements, but they also noted that the farmers only identified 54 percent of the projects on their farm that could deliver such improvements. A drawback of the reverse auction is that the farmers may well not have an advisor's knowledge of environmental investment options. However, again the farmers' knowledge would be expected to increase if further auctions took place. A future modification could be to produce a hybrid scheme where an advisor visited the farms and scored the investments, and this score, in turn, would contribute to the value-for-money score used to assess the bids. Alternatively, there is the opportunity to offer training courses to farmers to help close the knowledge gap.

Second, it was important to determine whether the capital investments made were ones that the farmers would have made without the grants. Surveys taken after the auction showed that 62 percent of farmers stated they "would have undertaken investments irrespective of whether they received funding from SWW." However, almost all respondents caveated this statement with responses like "only when, and if, alternative finance became available to fund it;" "not now and probably a number of years in the future;" and "to a lower standard than proposed in the bid." At the very least, these grants pushed forward the priority of capital investments that benefit the environment.

An improvement might be made in the way water quality enhancements were verified in the reverse-auction scenario by shifting from payments for action to payments for outcome. The problem with payments for action is that there is no guarantee the water quality in the river will increase from the changes to capital made on the farms. Without measuring the outcome of the changes, it is also impossible to know how much effect each capital investment had. While ongoing improvements in monitoring technology will help greatly here, in practice it is likely that a mixture of payments for action and outcomes will persist, in part to defray the risks to farmers inhibiting participation.

Key Lessons Learned

Several key lessons have been learned by this case study:

- Business-to-business investments can produce win-win outcomes—in this case, specifically benefiting the firm, the farms, water quality, associated biodiversity, and society.
- Increasing the information available to the investor increases the value those investments have to society.
- A reverse auction can be used to encourage the providers of environmental goods to reveal their costs to the investors.

- Assessing bids on a value-for-money basis helps ensure the best investments are made.

Conclusions

This chapter has provided an overview of the flexibility of PES designs across different geographical and administrative scales and both public and private sector involvement. Taking this information together shows the considerable potential that exists to refine current decision-making approaches so as to deliver economically efficient environmental enhancements.

Key References

Baker, David J., Stephen N. Freeman, Phil V. Grice, and Gavin M. Siriwardena. 2012. "Landscape-scale responses of birds to agri-environment management: A test of the English Environmental Stewardship scheme." *Journal of Applied Ecology* 49, no. 4:871–82. doi:10.1111/j.1365-2664.2012.02161.x.

Bateman, Ian, Matthew Agarwala, Amy Binner, Emma Coombes, Brett Day, Silvia Ferrini, Carlo Fezzi, Michael Hutchins, Andrew Lovett, and Paulette Posen. 2016. "Spatially explicit integrated modeling and economic valuation of climate driven land use change and its indirect effects." *Journal of Environmental Management* 181:172–84. doi:10.1016/j.jenvman.2016.06.020.

Bateman, Ian J., and Ben Balmford 2018. "Public funding for public goods: A post-Brexit perspective on principles for agricultural policy." *Land Use Policy* 79:293-300. https://authors.elsevier.com/sd/article/S0264837718308603.

Bateman, Ian J., Amy Binner, Brett H. Day, Carlo Fezzi, Alex Rusby, Greg Smith, and Ruth Welters. 2018. "A natural capital approach to integrating science, economics and policy within decision making: Public and private sector payments for ecosystem services." *LEEP Working Paper*, Land, Environment, Economics and Policy Institute (LEEP), University of Exeter Business School. http://www.exeter.ac.uk/leep/publications/workingpapers/.

Bateman, Ian J., Amii R. Harwood, Georgina M. Mace, Robert T. Watson, David J. Abson, Barnaby Andrews, Amy Binner et al. 2013. "Bringing ecosystem services into economic decision-making: Land use in the United Kingdom." *Science* 341, no. 6141:45–50.

Day, B., L. Couldrick, R. Welters, A. Inman, and G. Rickard. 2013. *Payment for Ecosystem Services Pilot Project: The Fowey River Improvement Auction*, Final Report to the Department for Environment, Food and Rural Affairs, LEEP (University of Exeter) and Westcountry Rivers Trust.

Defra. 2018. "Health and harmony: The future for food, farming and the environment in a Green Brexit." *Cm 9577*.https://www.gov.uk/government/consultations/the-future-for-food-farming-and-the-environment.

H. M. Government. 2018. *A Green Future: Our 25 Year Plan to Improve the Environment.* www.gov.uk/government/publications.

Millennium Ecosystem Assessment. 2005. *Ecosystems and Human Well-Being Synthesis.* Washington, DC: Island Press.

UK–NEA. 2011. *The UK National Ecosystem Assessment: Synthesis of the Key Findings.* UNEP-WCMC, Cambridge, UK. http://uknea.unep-wcmc.org/.

Caribbean: Implementing Successful Development Planning and Investment Strategies

Katie Arkema

Countries around the world are increasingly interested in accounting for the contribution of ecosystems to livelihoods and human well-being in their development plans and investments. Information about where and how ecosystems maintain water quality; sequester carbon; provide sources of sustenance; support tourism, agriculture, fisheries, and other sectors; and reduce the risk of communities to natural hazards can inform where to target investments in infrastructure and conservation. However, few examples exist in which estimates of natural capital have been used to inform the design of a sustainable development plan and subsequent investments. Here we share two case studies, one from Belize and one from The Bahamas, in which modeling of ecosystem services under alternative development scenarios was used to design integrated management plans. Through these case studies we show how an ecosystem services assessment can be embedded within an on-the-ground stakeholder engagement process to design development plans that not only are broadly supported, but that are being implemented through financing mechanisms and specific actions. Our results illustrate how an ecosystem services approach can (1) help decision makers consider diverse objectives from the start of a planning process, (2) make explicit the suppliers and beneficiaries of a country's natural assets, and (3) show how decisions made today will influence planning objectives into the future. Our results also highlight the central role of government in facilitating the participation of diverse institutions and stakeholders in the planning and decision-making process. We anticipate that Belize and The Bahamas will serve as examples for other countries that seek to achieve inclusive green growth

economies. The next important steps will be to monitor and adapt these plans to improve social, economic, and environmental outcomes.
....

In many countries around the world, major development projects are underway. An estimated US$57 trillion in global infrastructure development is anticipated by 2030, and the road network length is projected to increase 60 percent by 2050. As awareness grows about the role ecosystems play in supporting livelihoods and human well-being, national governments are pursuing development planning that accounts for the long-term health of people and the environment, as well as near-term economic goals.

Information about where and how ecosystems maintain water quality; sequester carbon; provide sources of sustenance; support tourism, agriculture, fisheries, and other sectors; and reduce the risk of communities to natural hazards can inform where to target investments in infrastructure and conservation. For example, assessments of natural capital can inform the finance mechanisms discussed in previous chapters by demonstrating where and how much restoration or conservation is needed to ensure the delivery of ecosystem services into the future.

Accounting for natural capital also helps to site development projects to leverage the benefits nature provides (and to avoid unnecessary losses of services). For example, considering the role that coastal and hillside forests play in preventing erosion and the importance of access to natural areas for tourism can help determine where new roads should be built and degraded roads maintained.

In this chapter we introduce two case studies from the Caribbean in which sustainable development planning was informed by an ecosystem services assessment. In particular, we explore how incorporating natural capital information in the planning phase helped to facilitate project financing during the implementation phase. In Belize, ecosystem services information was used to site ocean and coastal activities (e.g., dredging, fishing, oil exploration, ocean transportation, etc.) and to design the country's first Integrated Coastal Zone Management (ICZM) Plan. The plan has been used to inform investments in sustainable tourism development and nature-based coastal resilience projects. In The Bahamas, an ecosystem services approach to sustainable development planning is helping inform the scope of a loan from the Inter-American

Development Bank to the government of The Bahamas to restore mangrove forests for coastal resilience in the wake of two major hurricanes.

While Belize and The Bahamas are small Caribbean countries in terms of population and geographic size, the issues they face are relevant to countries around the world—even much larger countries like the United States and China, which have far larger coastal zones and more people living in harm's way. Globally, more than a third of the human population lies within 100 km of the ocean. The Bahamas and Belize highlight the role that coastal and marine ecosystems play in supporting economic development while enhancing coastal resilience from natural hazards in the face of global climate change.

Case 1. Integrated Coastal Zone Management and Investments in Development, Conservation, and Restoration in Belize

The Problem
Located along the eastern coast of Central America, Belize is known for its unique coastal and marine ecosystems. The highly productive coastal zone of Belize is home to a diversity of species, including endangered manatees and sea turtles, as well as 35 percent of the human population of Belize. Many shoreline communities rely on fishing for sustenance and livelihoods. World-renowned snorkeling and diving draw more than 800,000 tourists annually. In addition to attracting fishers and tourists, coral reefs, mangrove forests, and seagrass beds shield coastal communities from flooding and erosion. Despite the importance of Belize's marine environments, economic development of ocean and coastal sectors, including tourism, aquaculture, oil and gas, and marine transportation, threaten the very ecosystems that underlie the national economy and support human well-being.

To address the unstructured nature of economic activities in the coastal zone, and to ensure sustainable ecosystems for generations to come, the government of Belize embarked upon a massive effort to design the nation's first Integrated Coastal Zone Management (ICZM) Plan. The 2000 Coastal Zone Management Act called for a plan based on both science and local knowledge, which would be national in scope but emphasize regional differences and provide spatially explicit guidance about where and how to engage in ocean and coastal activities to achieve both conservation and development goals. The Belize Coastal Zone Management Authority and Institute (CZMAI) partnered with the Natural Capital Project (NatCap) in 2010 to use ecosystem service models to assess tourism, fisheries, and coastal protection objectives under several future scenarios for zoning ocean and coastal activities. The highly iterative and participatory process of stakeholder engagement and ecosystem modeling ultimately yielded a preferred zoning scheme for the ICZM plan that in turn helped to facilitate financing for investments in restoration and conservation of coastal and marine ecosystems. The plan was approved in 2016.

The Ecosystem Services

Coastal and marine ecosystems, such as coral reefs, mangrove forests, and seagrass beds, provide numerous ecosystem services to the Belizean people and visitors worldwide. Tourism is the single largest contributor to the country's economic growth, generating US$264.4 million in 2008 and more than a million visitors in 2014. The marine environment is a large part of what draws visitors. For example, tourists come to the country to experience the Belize Barrier Reef Reserve System and the Great Blue Hole, both of which are within a United Nations Educational, Scientific and Cultural Organization (UNESCO) world heritage site. In turn, their expenditures for recreational activities, (e.g., snorkeling SCUBA diving), food, lodging, entertainment, and other amenities support local economies.

Fishing is another benefit supported by coastal and marine ecosystems in Belize. In addition to fin fish, lobster, conch, and other harvested species, coral reefs, mangroves, and seagrasses provide important nursery and adult habitat for ecologically and economically important species. For example, the spiny lobster, which make up the largest share of the export fishery in Belize, recruits to mangroves and seagrasses from the open ocean. In the coastal vegetation, the lobsters grow to a juvenile life stage, eventually migrating offshore to seagrasses and coral reefs where they live and are caught as adults.

Coral reefs, mangrove forests, and seagrass beds also provide coastal protection from sea-level rise and storms. Through changes in depth caused by the barrier reef and the frictional effects of coral and vegetation, ecosystems extract wave energy, attenuate water flow, and trap sediments. Their effects on nearshore hydrodynamics and sediment transport can in turn reduce the impact of coastal erosion and flooding on shoreline communities. When coastal and marine ecosystems are degraded, through pollution, dredging, storm damage, and other stressors, their ability to provide fisheries, tourism, and coastal protection services suffers, as do the Belizeans and visitors relying on these services.

Ecosystem Service Beneficiaries

Coastal property owners—both single family homes and hotels and lodges—benefit from the ability of coral reefs, mangroves, and seagrass to reduce coastal flooding and stabilize shorelines. More than a third of the Belizean population living in the coastal zone benefits from risk reduction provided by ecosystems for public infrastructure and services (e.g., hospitals, schools, etc.). People involved in small-scale artisanal and subsistence fisheries and commercial fishing operators benefit from harvested species, such as spiny lobster and conch, as do other components of the fishing sector, including the processors, exporters, and their employees. Beneficiaries of healthy coastal ecosystems for tourism are numerous and include owners and employees of small lodges and ecotourism resorts, large resorts and hotels, restaurants, retailers, outdoor adventure operators, and other businesses that tourists frequent.

Ecosystem Service Suppliers

Coastal and marine ecosystem services are supplied by both public and private lands along the coast and within nearshore waters. As in many countries, coastal resources and public lands are managed by multiple agencies. Accounting for the role that ecosystems play in supporting different coastal sectors can enhance coordination among these agencies to ensure achievement of economic, environmental, and social objectives. For example, incorporating nursery habitat into the models for quantifying catch and revenue of spiny lobsters highlighted the economic importance of mangroves, which are managed by the department of forestry, to lobster, which are managed by the fisheries department. In addition to zoning and management of public areas, private landowners also supply ecosystem services. For example, a program in southern Belize was designed to reward waterfront property owners who invested in mangrove restoration and conservation for its multiple coastal resilience, carbon sequestration, and fisheries benefits. As another example, several resorts along the Placencia Peninsula, an area known for its beautiful beaches, have avoided construction of seawalls, which can lead to long-term erosion and changes in coastal geomorphology. Instead, owners would rather rebuild the resorts after a major hurricane than risk losing the primary resource—the beach—that ultimately supports tourism.

Design of the Country's First National Integrated Coastal Zone Management (ICZM) Plan

To minimize ecological degradation and better manage conflicting uses of the coastal zone, the government of Belize embarked on a process to develop the first national ICZM plan for the country. The approach combined stakeholder engagement, scenario development, and quantitative modeling of ecosystem services. The stakeholder engagement process involved scoping objectives, gathering information, and eliciting feedback through coastal advisory committees composed of local representatives from diverse sectors and interests, public consultations, and expert reviews. Scenario development involved synthesizing spatial data for eight ocean and coastal activities (development, transportation, fishing, agricultural runoff, tourism, dredging, aquaculture, and oil exploration) to develop one current (2010) and three future (2025) zoning scenarios based on stakeholder input, government reports, and existing and pending legislation. Spatial variation in ecosystem services in the current and future scenarios were quantified in biophysical and economic metrics using the tourism and recreation, coastal protection, and fisheries models in the InVEST software suite.

Through iteration of stakeholder engagement, scenario development, and modeling ecosystem services, the Belize Coastal Zone Management Authority and Institute (CZMAI), in collaboration with the Natural Capital Project, developed a preferred spatial plan for ICZM in Belize. Analysis of ecosystem services suggested that the preferred plan would lead to greater returns from coastal protection and tourism than outcomes from scenarios oriented toward achieving either conservation or development goals. The plan would also reduce impacts to coastal habitat and increase revenues from lobster fishing relative to current management. By accounting for spatial variation in the impacts of coastal and ocean activities on benefits ecosystems provide to people, the results from modeling ecosystem services allowed stakeholders and policymakers to make informed decisions about the extent and location of the zones of human use in the plan. Including outcomes in terms of ecosystem service supply and value (e.g., fisheries catch and revenue, coastal land protected from storms and avoided damages, and number of tourists and tourism expenditure) allowed for explicit consideration of multiple benefits from oceans and coasts that typically are evaluated separately in management decisions (fig. 16.1).

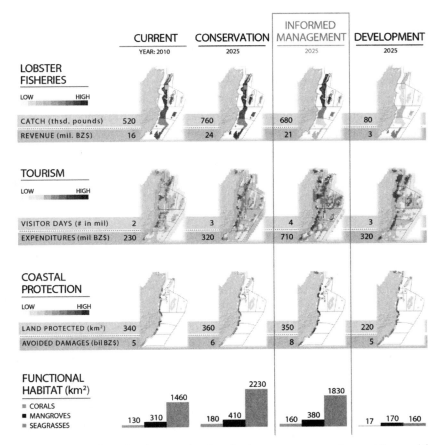

	CURRENT	CONSERVATION	INFORMED MANAGEMENT	DEVELOPMENT
	YEAR: 2010	2025	2025	2025

LOBSTER FISHERIES

LOW ▬▬▬ HIGH

	CURRENT	CONSERVATION	INFORMED MANAGEMENT	DEVELOPMENT
CATCH (thsd. pounds)	520	760	680	80
REVENUE (mil. BZ$)	16	24	21	3

TOURISM

LOW ▬▬▬ HIGH

	CURRENT	CONSERVATION	INFORMED MANAGEMENT	DEVELOPMENT
VISITOR DAYS (# in mil)	2	3	4	3
EXPENDITURES (mil BZ$)	230	320	710	320

COASTAL PROTECTION

LOW ▬▬▬ HIGH

	CURRENT	CONSERVATION	INFORMED MANAGEMENT	DEVELOPMENT
LAND PROTECTED (km²)	340	360	350	220
AVOIDED DAMAGES (bil BZ$)	5	6	8	5

FUNCTIONAL HABITAT (km²)

- CORALS
- MANGROVES
- SEAGRASSES

	CURRENT	CONSERVATION	INFORMED MANAGEMENT	DEVELOPMENT
CORALS	130	180	160	17
MANGROVES	310	410	380	170
SEAGRASSES	1460	2230	1830	160

Figure 16.1. Biophysical and economic values for three ecosystem services and the area of habitat capable of providing services under the current and three future scenarios for the ICZM Plan for Belize. This figure first appeared in the *Proceedings of the National Academies of Sciences* (Arkema et al. 2015).

How the Belize ICZM Plan Is Being Used to Inform Permitting and Investments

By accounting for ecosystem services, the Belize ICZM plan highlights how the country's coral reefs, mangrove forests, seagrasses, and other coastal and marine ecosystems play a critical role in national economic development and coastal resilience from natural disasters. This information about the economic and societal value of ecosystems helped to pave the way for financing conservation and restoration and investments in sustainable development. For example, the Inter-American Development Bank (IDB) is working with the Belize Tourism Board to finance sustainable tourism projects that are consistent with the ICZM Plan. The IDB has also invested in three shoreline stabilization projects in Toledo, Cay Caulker, and Corazal, which have considered nature-based project designs. Traditional approaches to coastal protection (e.g., seawall construction) is often under the jurisdiction of agencies responsible for major infrastructure projects. In the three nature-based coastal resilience projects in Belize, a suite of agencies is involved, including CZMAI and the Ministry of Tourism. Because natural and nature-based coastal resilience projects have the potential to meet multiple objectives, they often have the support of a suite of stakeholders and agencies and thus may also benefit from more diverse sources of financing.

Monitoring and Verification

Monitoring and verification are underway in Belize for both the ICZM Plan and the tourism and shoreline stabilization projects. The ICZM Plan will be revised over a four-year time period. To inform adaptive management of the coastal zone, CZMAI has partnered with scientists, nongovernmental organizations (NGOs), and students in sustainability science to design a monitoring protocol. CZMAI's Data Centre under the Marine Conservation & Climate Adaptation Project is responsible for managing an inventory of development sites and activities for the coastal areas within the nine planning regions. This inventory will serve to assess ICZM Plan compliance and monitor proper implementation of the development guidelines for each region within the four-year time period. In addition, NGOs such as Healthy Reefs for Healthy People are engaged in ecological monitoring that can be used to evaluate the ecological outcomes of the ICZM Plan. A regular report card put out every few years synthesizes field data on coral reef cover, fish density and biomass, and algal cover to track the health of the reef system over time. The shoreline stabilization projects are also being monitored to track changes in local hydrodynamics and sedimentation processes.

Effectiveness

With the ICZM plan approved in the middle of 2016 and the implementation of the shoreline stabilization projects still underway, it is too soon to know the efficacy of either the ICZM plan or the investments in shoreline protection.

However, since the formal approval of the ICZM plan by the government of Belize in the middle of 2016, CZMAI has completed an institutional assessment and identification of strategic actions toward strengthening the institutional structure and legislative framework for improved coastal zone management in Belize. CZMAI has also developed a "Short term Road Map" in support of coordinating the implementation of the Belize ICZM Plan. In addition, CZMAI has improved implementation of broad-based education and stakeholder engagement activities including, but not limited to, Coastal Awareness Week 2017, celebrated in partnership with the IDB under the theme "Business and the Environment: A Foundation for Sustainable Growth."

Key Lessons Learned

The Belize case study highlights several key lessons for assessing natural capital to inform coastal zone management and development. First, multiple rounds of stakeholder engagement, modeling of ecosystem services, and scenario development legitimizes the planning process and allows planners to incorporate both societal and scientific knowledge and solutions. Second, accounting for ecosystem services can inform the design of ICZM plans to enhance delivery of tourism, fisheries, and coastal protection benefits beyond that which would have otherwise resulted from stakeholder engagement alone. Third, including outcomes in terms of ecosystem service supply and value allow for explicit consideration of multiple benefits from oceans and coasts that typically are evaluated separately in management decisions. Fourth, an ecosystem services framework for coastal planning provides a common language through which multiple agencies can communicate and coordinate to achieve a shared vision.

In Belize the central government played a key role in enabling the use of natural capital information in development planning and investment. CZMAI, the coastal planning agency, facilitated stakeholder engagement at regional and local scales to ensure participation from the private sector and civil society. The participatory process was critical to designing alternative development scenarios and identifying critical ecosystem services. Coastal planners also worked closely with scientists to ensure the technical analysis directly informed decision making in real time. Ultimately, the ICZM Plan born from this process was designed to meet the requirements of the Coastal Act of 2000, which itself came to fruition after more than a decade of dedicated work on the part of NGOs, members of civil society, and key government officials.

Box 16.1

Valuing Nature in Myanmar as the Basis of Economic Development and Decision Making

Hanna Helsingen, Nirmal Bhagabati, and Sai Nay Won Myint

Myanmar at the Crossroads

Myanmar is at the crossroads of development: after years of isolation, the country started opening up in 2012, increasing both opportunities and threats to the country's people, economy, and natural capital. There are signs that the government wants to learn from the mistakes of other countries and pursue a more sustainable development pathway. Since an early commitment to a green economy by former president U Thein Sein in 2013, there is growing evidence of this policy commitment toward valuing nature and pursuing a green economy approach through Myanmar's new Sustainable Development Plan (MSDP).

Telling the Policy Story through Mapping of Myanmar's Natural Wealth

In 2015, the World Wide Fund for Nature (WWF), together with the Ministry of Natural Resources and Environmental Conservation, the Natural Capital Project, and Columbia University, carried out the first national natural capital assessment, based on a direct request from the then-president of Myanmar. By incorporating climate risks into the assessment, it also became evident that the country's natural capital was both under threat from climate change and crucial to climate change adaptation. This assessment has become an important narrative for guiding development policies and plans toward better valuing nature in Myanmar.

New Policy and Planning Documents in Place

Myanmar is creating and updating a number of plans and policies, reflecting an increased understanding of the importance of natural capital and the use of information about its value. The government hopes that accounting for the value of nature will drive increased public and private investments toward conservation and sustainability, better planning and design of large-scale infrastructure investments, and increased benefits to local communities. As an example, the National Environmental Policy now emphasizes the need to understand and use the value of nature as a basis for economic and social development.

The 2018 Myanmar Sustainable Development Plan (MSDP) recognizes the natural environment as the foundation for all economic and social development. It emphasizes the need to incorporate the value of nature into infrastructure planning and design, national accounting systems, and green economy planning to support low-carbon and climate-resilient development. The most important change the MSDP could bring is that private and public investments are aligned with its objectives and actively support its implementation. This would mean that investments are no longer to be assessed solely on economic potential but also incorporate social and environmental costs and benefits.

Myanmar's new investment law (2016) also highlights this by stating that all investment must benefit the economy and people, and do no harm to the environment. Its investment rules direct environmental impacts to be assessed as part of the upstream decision-making process as part of the investment proposal assessment criteria. There is promise in the level of ambition in these documents. The biggest challenge now lies in their implementation and application.

A Path Forward

Additional work is needed to ensure that information is (1) described in ways that highlight the losses at stake from not incorporating the value of nature, and (2) provided at the right time for decision making. This will be crucial for ensuring that policies and information translate into a greener economy in Myanmar, where the value of nature is at the core of how the country develops.

Key References

Government of Myanmar. 2018. Climate Change Strategy and Action Plan (draft). Naypyitaw.
———. 2018. Draft Green Economy Policy Framework. Naypyitaw.
———. 2018. Myanmar Sustainable Development Plan. Naypyitaw. http://themimu.info/sites /themimu.info/files/documents/Core_Doc_Myanmar_Sustainable_Development_Plan_2018 _-_2030_Aug2018.pdf.
———. 2018. *National Environmental Policy* (Final). Naypyitaw.
Mandle, Lisa, S. Wolny, P. Hamel, H. Helsingen, N. Bhagabati, and A. Dixon. 2016. "Natural connections: How natural capital supports Myanmar's people and economy." Yangon: WWF– Myanmar. http://www.myanmarnaturalcapital.org.
World Bank. 2015. "Myanmar—Post-disaster needs assessment of floods and landslides: July– September 2015." Washington, DC: World Bank Group.

Case 2. Development Planning for Andros Island in The Bahamas

The Problem

The Island of Andros lies forty miles to the west of Nassau, the capital of The Bahamas. Encompassing a land area greater than all other 700 Bahamian islands combined, Andros remains largely undeveloped. Vast mangrove and coppice forests, the third largest coral reef in the world, seagrass beds, sand flats, and a concentrated system of blue holes (unique, roughly circular steep-walled carbonate depressions named for their deep blue waters) support the country's commercial and sport-fishing industries, nature-based tourism activities, agriculture, and freshwater resources. There has been a concerted effort to provide the basic elements of human development to every island of the Bahamian archipelago. However, many parts of Andros, due to the sparseness of the population, which impacts scale efficiencies, require better infrastructure and training programs to support themselves and ensure the well-being of generations to come. The central challenge confronting the government of The Bahamas is to make investments that support sustainable economic development and educational opportunities, while also harnessing the island's wealth of natural assets and without sacrificing the ecosystems that underlie its economy and sustain the well-being of its citizens.

To address this challenge, the Office of the Prime Minister (OPM), with support from the Inter-American Development Bank (IDB), engaged in an innovative process to design a Sustainable Development Master Plan for Andros Island. The goal was to identify public and private investment opportunities, policy recommendations, zoning guidelines, and other management actions to guide the sustainable development of the island over the next twenty-five years. In consultation with Androsians, government agencies, NGOs, and other stakeholders, OPM worked with the Natural Capital Project (NatCap) to design a preferred island-wide future scenario depicting where human activities and a suite of environmental and social objectives could be harmonized. In early 2017, the planning process delivered a community-supported vision within which to embed specific development projects, policies, and investments.

The Ecosystem Services

The ecosystems of The Bahamas deliver a suite of important benefits. As one of the more pristine islands, the environment of Andros is particularly unique. Andros has one of the highest densities of blue holes in the world. In addition to attracting tourists, these blue holes and other fascinating aspects of Andros' geology are research sites for scientists from around the world. The extensive sandflats and freshwater creeks provide critical nursery habitat for bonefish, a lucrative recreational fishery. Land crabs, a local delicacy and source of income generation, also use multiple coastal and marine habitats, with females laying their eggs in the nearshore waters, the early life stages occurring in the ocean, and juveniles and adults living in holes in the floor of the coastal forest.

Like many small island-developing states, The Bahamas is particularly vulnerable to natural disasters. Indeed, in the last four years several hurricanes have hit the islands. Hurricane Matthew, in 2016, impacted the entire island chain including Andros Island. The results were approximately US$600 million in damages across the country. Degraded seawalls and coastal structures highlight the challenge of investing in built infrastructure that requires regular maintenance. This challenge is amplified in a country of more than 700 islands, with the majority of its financial and human capital concentrated on a single island (New Providence). As an alternative to traditional shoreline stabilization structures, the coral reefs, mangroves, coppice forests, and seagrass beds create a three-dimensional structure that traps sediments and attenuates waves.

Ecosystem Service Beneficiaries

All of the main settlements on Andros are coastal, which means that nearly the entire population of the island benefits to some extent from the coastal resilience provided by the barrier reef, coastal forests of mangrove and coppice, beach grasses, and underwater seagrasses. In particular, coastal hazard analyses conducted with the InVEST Coastal Vulnerability model, combined with census data for Andros, show that 50 percent of the population in the most exposed areas benefits from risk reduction provided by coastal and marine ecosystems. Other beneficiaries of sustainable development include the more than 2,500 fishermen for spiny lobster, conch, grouper, snapper, and other fish living on the coral reef, across the Great Bahama Bank, and in the nearshore seagrass and sandflats. The bonefishing sector on Andros is particularly lucrative with sport fishermen spending several thousand dollars for a week of lodging on Andros and a local guide to take them to the best spots by boat. Finally, with the land crabs selling for US$5–8 apiece, many an islander spoke of how the local and nearby markets for crab "put my kids through college."

Ecosystem Service Suppliers

Much of Andros Island is crown land—that is, owned by the government. In addition, many ecotourism lodges and small resorts dot the east coast of the island, which is the most populated area. Several of the owners have developed eco-conscious businesses both out of personal interests and because of the economic return from guests who are seeking a sustainable tourism experience. For example, Small Hope Bay has engaged in dune restoration for coastal resilience and put up signage explaining the benefits of the dunes to guests. Another resort, Kamalame Cay, caters to a high-end clientele interested in conservation, its cottages open to the sea, taking advantage of the ocean breeze in place of air-conditioning. The owner of this resort lives nearby in a home he built for his family tucked inside the mangrove forest to protect it from storms.

Design of Sustainable Development Master Plan for Andros Island

The government of The Bahamas worked with stakeholders, agencies, and NGOs to design a Sustainable Development Master Plan for Andros Island. The goal of the plan was to identify public and private investment opportunities, policy recommendations, and zoning guidelines that would improve the livelihoods of Androsians, provide them with training and educational opportunities, improve access to tourists and employment, enhance coastal resilience, and safeguard ecosystems that underlie human well-being. The approach was similar to the one used in Belize in that it included stakeholder engagement, scenario development, and quantitative modeling of ecosystem services. However, in The Bahamas, because the financial support for the planning process was from a multilateral development bank, and the central government (the Office of the Prime Minister) led the project, there was an even greater focus on development projects and the importance of ecosystems to human well-being and economic benefits.

Through an extensive engagement process, which involved stakeholders communicating their preferences for the future using hand-drawn and printed maps, the planning team created spatial data layers reflecting the current situation of human activities and four future development scenarios. The *Business as Usual* scenario represents a future similar to the current situation with little investment in new infrastructure, educational opportunities, or development. The *Conservation* scenario gives priority to ecosystem health and protection of habitats and species rather than economic development. For example, this scenario includes the ratification of a national park for the Andros barrier reef, but no new coastal development. The *Sustainable Prosperity* scenario blends human development and conservation goals by investing in critical infrastructure and education to achieve a nature-based economy that can be sustained over time. Examples of activities include daily ferries from Nassau, small and midsized Bahamian-owned businesses (e.g., hotels, processing factories for local goods), community agriculture, and mangrove restoration, as both a natural means of shoreline protection from storms and a habitat for lobster. The *Intensive Development* scenario gives priority to high impact economic development rather than ecosystem health and protection of habitats and species.

Example activities include construction of a cruise ship port in North Andros; large, energy intensive resorts and luxury housing developments; expanded mining activities; and seawalls along the entire east coast of the island.

Through iteration of stakeholder engagement, scenario development, and modeling ecosystem services, the OPM, in collaboration with Androsians, the University of The Bahamas, and NatCap, designed the Sustainable Prosperity scenario as the preferred scenario for sustainable development of Andros Island. A consulting firm then worked with the planning team and stakeholders to identify a suite of priority projects that would realize this scenario over a five-, ten-, and fifteen-year time frame. Importantly, using an ecosystem services approach and participatory mapping showed that the citizens of Andros were less eager about massive new development projects that could pose severe risk to ecosystems and, instead, wanted investments in degraded infrastructure like roads or processing plants for local goods within existing settlements that would allow them to better access—and safeguard—the wealth of natural resources on the island.

How the Sustainable Development Master Plan Is Being Used to Inform Investment Decisions

Accounting for ecosystem services in the design of the Andros Sustainable Development Master Plan has helped to prioritize the financing of ecosystem restoration and conservation of coastal and marine ecosystems in The Bahamas. From a conservation perspective, the plan has provided the central government with a framework against which to evaluate—at least informally—several development proposals. For example, proposals for extensive mining in and around Andros (especially related to aragonite) and for cruise ship development have been made. These kinds of activities and investments were included in the Intensive Development scenario for the future, which the ecosystem service models suggested would lead to decreases in tourism, fisheries, and coastal protection-related benefits, and which Androsians rejected in favor of the Sustainable Prosperity scenario. The plan includes an explanation and the results of the ecosystem services assessment across alternative futures. Thus, the plan provides quantitative information that stakeholders can point to during the evaluation of proposed infrastructure projects to show the potential costs to the environmental and human well-being of the island, and to communicate about stakeholder consensus for the future of the island.

Accounting for ecosystem services in the design of the master plan also helped finance ecosystem restoration. In particular, the analysis for the plan highlighted the potentially negative consequences of traditional approaches to shoreline protection (e.g., seawalls) for the long-term health of fisheries, the tourism industry, and shoreline erosion. As a result of this work, the government of The Bahamas and their development financing partners are considering mechanisms to fund several different coastal management projects across the islands. Additionally, because of the ecosystem service analysis for the Andros Master Plan, the IDB and government of The Bahamas are interested in using this island as a site to demonstrate the effectiveness of natural infrastructure for shoreline stabilization. The project includes US$3 million for restoration of ecosystems such as mangroves. In addition to the coastal resilience benefits, the cobenefits (e.g., fisheries, tourism) of natural approaches to shoreline stabilization and flood protection are essential to the economy and livelihoods on Andros.

Monitoring and Verification

No formal protocols have been developed for monitoring the Sustainable Development Master Plan for Andros. However, over the past several years The Bahamas has been engaged in a national development-planning initiative called Vision 2040. As a part of this effort, the Development Planning unit in the Office of the Prime Minster has integrated the United Nations Sustainable Development Goals (SDGs) into the planning framework for Vision 2040, as well as the monitoring and evaluation framework to track progress. There is also potential for applying the SDGs and their corresponding targets at the island scale to monitor master plans. In the case of the Andros Sustainable Development Master Plan, the ecosystem service objectives that the Office of the Prime Minister, NatCap, the IDB, and the University of The Bahamas used to explore the alternative scenarios proposed by stakeholders map nicely onto several of the SDGs. The goals that align well with the Andros plan include Life Below Water (SDG14), No Poverty (SDG1), Zero Hunger (SDG2), Good Health and Wellbeing (SDG3), Climate Action (SDG13), and Sustainable Cities and Communities (SDG11).

Effectiveness

With the Andros Sustainable Development Master Plan, recently approved in 2017, and a new partnership operation for mangrove restoration still in development, it is too early to know the effectiveness of the plan for achieving its long-term ecological, economic, and environmental goals. However, anecdotally the planning process has already been effective as a way for Androsians to identify and communicate what they want for their future. For example, the central government receives major development proposals and inquiries from national and international corporations with interests in Andros (e.g., timber, cruise ship, mining) and the Andros Master Plan provides a basis against which to review these proposals. In any country, but especially in an archipelago, it is often difficult for local communities to communicate with the central government about their priorities. With a road map in hand, such as the Andros Sustainable Development Master Plan, Bahamians are better equipped to say whether proposed projects align with the overarching vision and specific types of investments they desire for the future of their island. In addition, accounting for ecosystem services in development planning changed the conversation around investments in coastal infrastructure for disaster risk reduction. As a result, engineers in the Ministry of Works, Housing, and Urban Development are increasingly considering alternative natural and nature-based approaches such as mangrove restoration for coastal resilience.

Key Lessons Learned

The Bahamas case study highlights several key lessons from assessing natural capital to inform coastal zone management and development. First, full and repeated engagement among policymakers, scientists, and stakeholders in building alternative visions of the future focuses and legitimizes the planning process. Second, measuring and comparing changes in the benefits of nature across scenarios helps people identify shared goals and understand trade-offs. Third, hand-drawn and printed maps reflecting stakeholder recommendations and large posters illustrating changes in ecosystem services can serve as "boundary objects" during a sustainable development process, around which stakeholders, scientists, and policymakers can come together and share ideas. Fourth, ecosystem services can be used to evaluate the impact and benefit of a proposed master plan in economic, environmental, and social metrics, thus encouraging consideration of all three aspects of sustainable development together, throughout a planning process.

Similar to the work in Belize, the central government in The Bahamas played an important role in facilitating the science-policy process. In particular, through its convening role the Office of the Prime Minster was able to leverage both local knowledge from the stakeholder-engagement process and technical analysis from the science team to inform coastal development planning and investments. The Inter-American Development Bank also played a critical role by funding a planning process that combined stakeholder engagement, scenario development, and the assessment of natural capital. And finally, participation by other institutions, such as The University of The Bahamas, environmental NGOs, and the private sector was critical for maintaining continuity after a change in government leadership.

Conclusions

Together, the Belize and The Bahamas case studies illustrate several important insights for natural capital assessment and development-planning efforts underway in other countries. First, stakeholder participation is critical for legitimizing the process, fostering buy-ins of planning objectives and outcomes, and ensuring realistic proposals and investments. In both Belize and The Bahamas, stakeholders played a key role in filling knowledge gaps about where activities occur in the coastal zone and designing alternative development scenarios. However, stakeholder engagement is time and resource intensive. The central government in both cases was an important convener of public meetings and regional planning bodies. Yet it also partnered with diverse groups and institutions to support the planning process. In Belize, local leaders were chosen by their peers to facilitate coastal advisory committee meetings. In The Bahamas, local government officials and a consulting firm with extensive experience in the region conducted the stakeholder engagement. Other institutions such as environmental NGOs (e.g., World Wildlife Fund, Healthy Reefs, and The Nature Conservancy), universities, and multilateral development banks provided the continuity in the planning process when elections led to a change in the national leadership.

Incorporating multiple ecosystem services in development planning helps to encourage the participation of diverse actors with a variety of goals. Coastal planning initiatives in both Belize and The Bahamas integrated across agencies and industries such as hazard mitigation, emergency response, fisheries, agriculture, and tourism. Including technical assessment of natural capital as it supports these industries was crucial for elevating the importance of environmental sustainability in national development planning. For example, in The Bahamas, estimating the financial contribution of nursery habitat on Andros for economic returns from the lobster fishery highlighted how development decisions for a particular island can have economic impacts at a national scale for an important sector such as fishing.

Finally, these cases highlight the importance of finance mechanisms for implementing integrated coastal development plans. In The Bahamas, the Inter-American Development Bank was involved from the start of the process. Thus, when the opportunity arose to propose a loan for coastal management following two major hurricanes, the government of The Bahamas and the Bank were better positioned to explore investments in innovative nature-based approaches to reducing risks from coastal hazards that would also support the sustainability of other critical ecosystem services. There is a need for other forms of partnerships to ensure the sustainability of this work, particularly for small island-developing states, many of which are highly indebted.

Key References

Arkema, K. K., and M. Ruckelshaus. 2017. "Transdisciplinary research for conservation and sustainable development planning in the Caribbean." In *Conservation in the Anthropocene Ocean: Interdisciplinary Science in Support of Nature and People.* P. Levin and M. Poe, eds. San Diego, CA: Elsevier.

Arkema, Katie K., Gregory Verutes, Joanna R. Bernhardt, Chantalle Clarke, Samir Rosado, Maritza Canto, Spencer A. Wood et al. 2015. "Assessing habitat risk from human activities to inform coastal and marine spatial planning: A demonstration in Belize." *Environmental Research Letters* 9, no. 11:114016.

Arkema, Katie K., Gregory M. Verutes, Spencer A. Wood, Chantalle Clarke-Samuels, Samir Rosado, Maritza Canto, Amy Rosenthal et al. 2015. "Embedding ecosystem services in coastal planning leads to better outcomes for people and nature." *Proceedings of the National Academy of Sciences* 112, no. 24:7390–95.

Coastal Zone Management Authority and Institute (CZMAI). 2016. *Belize Integrated Coastal Zone Management Plan.* Belize City, Belize: Coastal Zone Management Authority and Institute. https://www.coastalzonebelize.org/wp-content/uploads/2015/08/BELIZE-Integrated-Coastal-Zone-Management-Plan.pdf.

Dobbs, R., H. Pohl, D. Y. Lin et al. 2013. *Infrastructure Productivity: How to Save $1 Trillion a Year.* Seoul: McKinsey & Company.

Dulac, J. 2013. *Global Land Transport Infrastructure Requirements.* Paris: International Energy Agency.

Sustainable Development Master Plan for Andros Island. February 2017. https://www.dropbox.com/s/dbkyw0lzvrbd6qk/AMPpercent20FINALpercent20VERSIONpercent20FEBpercent202017.pdf?dl=0.

Verutes, Gregory M., Katie K. Arkema, Chantalle Clarke-Samuels, Spencer A. Wood, Amy Rosenthal, Samir Rosado, Maritza Canto, Nadia Bood, and Mary Ruckelshaus. 2017. "Integrated planning that safeguards ecosystems and balances multiple objectives in coastal Belize." *International Journal of Biodiversity Science, Ecosystem Services & Management* 13, no. 3:1–17.

Cities: Incorporating Natural Capital into Urban Planning

Perrine Hamel, François Mancebo, Clément Feger, and Stéphanie Hamel

As humanity is becoming an ever more urban species, the role of cities on the global environmental agenda has never been as important. Mainstreaming natural capital in the field of urban planning can benefit both urban populations and the natural environment. Therefore, planners and decision makers need to understand when, how, and what type of nature-based solutions can address the specific challenges faced by their city. In this chapter, we present three case studies illustrating how nature's benefits have been incorporated into urban planning decisions. The cases address contexts as varied as urban agriculture policy and strategic urban planning, in the French cities of Nantes and Bordeaux, and a citywide integrated urban water management strategy in Melbourne, Australia. They show that urban green space provides multiple benefits, including efficient water management, food production, recreation opportunities, and the protection of biodiversity. We identify key opportunities and barriers for incorporating urban nature in planning, in particular related to participatory approaches and the integration of bottom-up initiatives into formal urban planning processes, which help legitimize the implementation of nature-based solutions. Management challenges include the selection of the portfolios of nature-based solutions that are adapted to the social, ecological, and technical attributes of urban systems. We conclude with a discussion of the mechanisms—top-down and bottom-up—that are available and proven for implementing and maintaining nature-based solutions.

Our world is increasingly urban. By 2030, urban areas are projected to be three times larger than in 2000, with a large variability in growth rates across continents (Seto et al. 2012). Urban growth intensifies the pressure on other land uses, such as for agriculture, forestry, recreation, and biodiversity conservation. For example, it is estimated that by 2030 urbanization will drive the loss of up to one-third of croplands in some African and Asian countries (d'Amour et al. 2017), and up to 7 percent of key biodiversity hotspots in some regions (Seto et al. 2012). Better understanding of urban systems is thus critical to help move toward a more sustainable, livable, and equitable future.

Cities depend crucially on ecosystems for security in food, water, climate, and health. They also face phenomenal challenges associated with their growth, social discontent, and natural hazards exacerbated by climate change such as floods, droughts, and heat waves. To address these challenges, engineers and planners are turning to new solutions, which include infrastructure relying on natural ecosystems. In parallel, many citizen initiatives are promoting a return of nature to the urban life. These include, for example, the creation of community gardens, installation of beehives or birdhouses, and the organization of planting days in urban parks. Thus, at various scales—neighborhoods, cities, planet—nature-based solutions offer promising prospects to manage urban systems.

Nature-based solutions is an umbrella term for a range of interventions that rely on natural systems to address societal challenges such as climate change, environmental degradation, natural resource scarcity, and natural disasters. The term was coined in the late 2000s in the context of climate adaptation. It has since broadened to encompass all types of win-win solutions that use nature's benefits—or ecosystem services—to solve these challenges, while also providing additional well-being or biodiversity benefits. Solutions can take various forms and spatial scales, from green roofs to insect hotels, urban farming, or the development of green corridors at the regional scale.

In the following sections, we illustrate how nature-based solutions have made their way into the urban planning process in three contexts: the implementation of a stormwater management strategy in Melbourne, Australia; a participatory process for urban planning in Bordeaux, France; and the spontaneous development of community gardens in Nantes, France. The case study from the San Francisco, California, Bay Area (chapter 14) also provides an example of nature-based solutions to enhance urban resilience to sea-level rise. Our chapter concludes with a synthesis of the lessons from the case studies and a discussion of the opportunities and barriers to incorporate nature-based solutions into urban planning.

Case 1. Urban Water Management in Melbourne, Australia

The Problem

The city of Melbourne has been recognized internationally for its record of innovation in the field of integrated urban water management. To address the multiple types of hazards it faces, including droughts, floods, and heat waves, the city has developed and implemented integrated water management plans that include nature-based solutions. This case study describes the experience of the Yarra Ranges Council, a municipality in the Greater Melbourne Area that has pioneered innovative approaches to stormwater management (fig. 17.1).

The strategies adopted by the Yarra Ranges Council were informed by the lessons from a ten-year research effort called the Little Stringybark Creek Project (https://urbanstreams.net/lsc/). Codesigned by the metropolitan water agency, Melbourne Water, and several research groups, the project partnership aimed to promote, test, and implement novel stormwater control measures at catchment scale.

The Ecosystem Services

Stormwater running off impervious areas degrades the quality of receiving waters and enhances flood risk. To avoid these problems, a suite of stormwater control measures can be implemented, including engineered stormwater control measures such as biofilters, swales, rainwater tanks, and green roofs (fig. 17.2). Such solutions help retain stormwater at the source, with the water being later infiltrated, evaporated, or repurposed. In addition, protecting and restoring forested areas, or setting a minimum of pervious areas in residential developments, can also increase stormwater retention.

In addition to the primary stormwater objectives, there are additional benefits of these stormwater control measures. Nature-based solutions, such as biofilters or buffer strips, help recharge groundwater, provide habitat for species, improve aesthetics, and promote recreation. Another important benefit, which indirectly promotes ecosystem health, is the reuse of stormwater stored in rainwater tanks: harvested water can be used for irrigation or for internal purposes such as laundry or toilet flushing, thereby potentially reducing water extraction from ecosystems. Using harvested stormwater in place of water mains across the municipality also reduces the demand on the natural rivers that are the source of most of the city's water supply. This helps to ensure there is enough water for both domestic use and environmental flows that support healthy aquatic ecosystems.

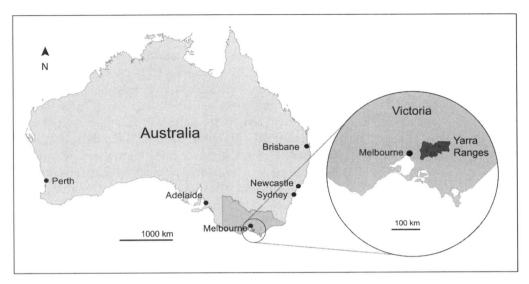

Figure 17.1. Location of Yarra Ranges Council in the Melbourne metropolitan area, Australia.

Figure 17.2. Examples of nature-based stormwater control measures implemented in the Council (Photo credit: P. Jeschke and S. Hamel.)

Ecosystem Service Beneficiaries

The use of nature-based solutions for stormwater management has multiple beneficiaries. First, it helps the Yarra Range Council reach its targets for urban water management. The Healthy Waterways strategy sets targets for impervious areas and stormwater harvesting for each metropolitan catchment as defined by Melbourne Water, the metropolitan water agency. Throughout its catchment, the Council aims to achieve the same Best Practice Environmental Management objectives originally only enforceable in residential subdivisions.

Second, it benefits local residents who can enjoy higher waterway quality and associated biodiversity as well as enhanced aesthetic quality for large-scale projects such as a sediment pond or wetland. Particular volunteer groups placing a high value on local creek systems have shown a great interest in nature-based solutions.

Finally, the implementation of nature-based solutions has benefits for education and research: multiple universities are studying the effect of the systems on water quality, and some educational programs were developed to raise awareness of the multiple benefits of nature-based solutions for urban water management. The political and academic value of the project also attracted funding to support the program, some of which benefited local residents. In fact, economic incentives were offered to landowners at different points in time to install raingardens (biofilters) or other stormwater control measures on their properties.

Ecosystem Service Suppliers

The suppliers of the service are the Council, which owns open space and right-of-ways where stormwater control measures were implemented, as well as private landowners, who are incentivized to implement similar solutions on their properties.

Terms of the Exchange: Quid Pro Quo

Under the State Planning Provisions' regulations, new residential subdivisions need to meet stormwater-retention targets related to water quality (retaining 80 percent of the annual loads of total suspended solid, 45 percent of total nitrogen and 45 percent of total phosphorous), and runoff volume. The Council is responsible for enforcing these regulations for new developments.

To this end, the Integrated Water Management Plan (IWMP), implemented by the Council, and the draft Water Sensitive Urban Design (WSUD) implementation plan, currently being codeveloped by two of the Council's departments, have two goals: to promote passive treatment, where possible, for all Council infrastructure; and to provide guidance on the implementation of stormwater control measures in high value priority catchments characterized by low Directly Connected Imperviousness (DCI). DCI, sometimes called effective imperviousness, is an indicator of urban stormwater impacts defined as the amount of impervious surfaces directly connected to streams by stormwater pipes and has been established as a primary driver of degradation of ecological health in urban waterways

There is currently no state policy that requires retrofit of catchments with nature-based solutions, meaning that provisions for catchments retrofit in the Council's IWMP and WSUD implementation plan are only considered as guidance documents. Additional water quality protection measures are implemented by other Council departments: for example, the Biodiversity Conservation and Parks and Bushland programs aim to protect and restore forested areas, while the Council's planning scheme stipulates that a standard of 20 percent of pervious surface will facilitate on-site stormwater infiltration in small residential developments in the absence of other conditions specified in the zone.

Mechanism for Transfer of Value

Several financial streams supported project implementation. Melbourne Water and the associated research program contributed financially to the implementation of stormwater retention measures, through both subsidies to the Council and direct incentives to private stakeholders. Financial compensations were offered to residents during the implementation of the Little Stringybark Creek project, for example, compensations for the installation of a leaky tank system on their properties. Leaky tank systems are rainwater harvesting systems in which an outlet connection is set to leak to an outdoor raingarden or similar feature to allow a constant detention volume to be stored. In addition, rate payers contribute to the stormwater management budget of the Council, which directly funds the implementation of stormwater retention measures. The maintenance side of the WSUD program, including a dedicated WSUD maintenance officer, associated maintenance budget, and other contributions to the infrastructure cost, were provided as part of this budget.

Monitoring and Verification

The Council keeps track of all public measures that contribute to the stormwater retention strategy. Through a regular audit undertaken every three years, the performance of all assets is quantified. Monthly inspections of each asset are also performed by the Council's WSUD maintenance officer to trigger necessary maintenance or retrofitting activities. All Council's assets and their conditions are kept in an internal asset mapping and registration system.

The research program associated with the Council has monitored multiple systems in the Yarra Ranges Council, to better understand the effectiveness of stormwater retention measures. Key indicators such as outflow volume and water quality were measured for these systems to improve future designs and retrofit existing low-performance systems.

Effectiveness

The effectiveness of the program can be assessed through three lenses: implementation of stormwater retention measures (and compliance with regulations), communication, and research objectives. Stormwater retention measures are effective in meeting regulatory targets in new residential subdivisions, where the best practices described above are achieved. In the catchment-scale experiment in Little Stringybark Creek, DCI rates have decreased from 6 percent to below 2 percent (relative to the total catchment area) in one area. No catchment-scale measurement is currently available for other retrofit efforts in other private or public areas.

From a governance perspective, the program helped identify key barriers and opportunities to stormwater retention implementation. For example, the partnership with a research program helped build relationships and credibility with landowners and facilitated the implementation of stormwater control measures on private land. The return on investment of stormwater control measures, including flooding damage reductions or ecological value, has limited data to date within the Council, and the implementation of the plans will allow further research in this area.

Key Lessons Learned

The Melbourne experience provided the following lessons for implementing nature-based solutions:

- Stormwater regulations represented a strong driver of the implementation of nature-based solutions where they apply. They motivated the design and implementation of an Integrated Water Management Plan and WSUD implementation plan based on the Best Practice Environmental Management Guidelines formulated in the state policy that included nature-based solutions.

- Additional cobenefits of nature-based solutions include irrigation water, internal use, and recreation, meaning that such solutions are increasingly considered in projects such as township improvements and master planning.
- The partnership among the metropolitan water agency, research groups, and the Council was critical to improving the design of engineered solutions and building trust for private landowners to take part in the program.
- Grants programs such as the Melbourne Water–funded Living Rivers program is vital for the ongoing implementation of the Council's WSUD program as it prompted the recognition of joint responsibility to deliver positive ecological impacts for the waterways.
- Synergies between the WSUD program and other sustainability, biodiversity conservation, and urban planning programs in a Council with high-value habitats are likely contributing to the high awareness level and investment toward nature-based solutions.

Case 2. Incorporating Ecosystem Services in Urban Development and Planning in Bordeaux, France

The Problem

The metropolis of Bordeaux, hereafter "Bordeaux" or "the city," consists of twenty-eight municipalities. It is located in the southwest of France, near the Atlantic coast (fig. 17.3). In addition to large areas of built-up urban land, which include the historical center, the peri-urban landscape also comprises small-scale agricultural land and wine estates; parts of the Landes forest in the northwest; large wetlands areas along the Garonne River; as well as rich biodiversity corridors along smaller rivers (e.g., Jalles-de-Blanquefort). The city is currently expanding and is expected to reach one million inhabitants by 2030 (from ~770,000 today). This raises important challenges regarding the ability of the city to ensure long-term protection of biodiversity and ecosystem services.

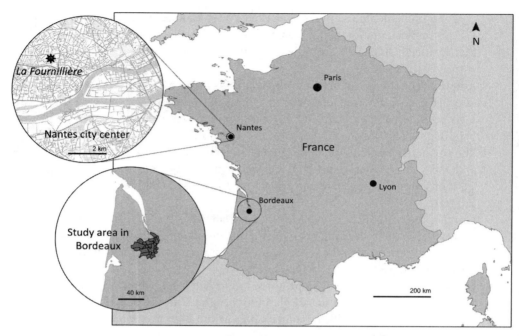

Figure 17.3. Location of Bordeaux (case study 2) and Nantes (case study 3) in France. Nantes city map layer from Stamen Toner Lite/Open Street Map.

The metropolitan authorities have carried out prospective studies since 2011 to better incorporate nature protection into landscape planning and management. The project described in this case study contributed to these reflections, using ecosystem services modeling and participatory methods. These involved public and private stakeholders participating in natural areas management (departments of the Bordeaux Metropolis: Nature, Water, Urban Planning, etc.), as well as the local water company, the agricultural chamber, the Regional Council, and two local conservation nongovernmental organizations (NGOs). The objective was to evaluate the status of Bordeaux's ecosystem services and to assess changes associated with alternative urban planning scenarios. The research was supported by the local water company, Lyonnaise des Eaux (now Suez Water). The company's motivations were to understand better its own impacts and dependence on ecosystem services, and also to open a dialogue with the metropolitan authorities and other stakeholders on green infrastructure development.

The Ecosystem Services

Focal ecosystem services were selected by the research team based on several interviews with the metropolis authorities as well as local stakeholders involved in natural areas and land-use management.

The regulation of water yield by the suburban areas' ecosystems was an important service to evaluate as Bordeaux is at significant risk of flooding. Across the urban development scenarios explored, the ability of ecosystems to regulate water flows would decrease by an estimated 8.8 percent (conservation scenario) to 10.5 percent (current urban plan). Projections for other water regulation services were made with the InVEST (Integrated Valuation of Ecosystem Services and Tradeoffs) models (www.naturalcapitalproject.org/invest). These included water quality regulation services (nitrogen and phosphorous retention), which were found to be on a degrading trend since 1990 and in all future scenarios considered. These services play an important role given the presence of agricultural activities, whose runoff increases the risks of algal blooms.

Three other important ecosystem services were assessed—carbon storage for global climate regulation, nature-based recreation, and food provisioning—along with habitat quality for biodiversity. Part of the Landes forest, which lies northwest of Bordeaux, contributes greatly to the city's carbon storage capacity and global climate regulation. Recreation ecosystem services were measured as "green areas" accessible to inhabitants living in the surroundings. Agricultural productivity was assessed as a proxy for food provisioning existing within the municipality's peri-urban areas. This has a particular relevance given the role of small-scale agriculture in the metropolis' prospective studies and the importance of vineyards in Bordeaux's identity. Finally, habitat degradation was also assessed as Bordeaux's natural areas are home to a rich and threatened biodiversity: amphibians, birds, and endangered reptiles and mammals, such as the European pond turtle or the European bison.

Ecosystem Service Beneficiaries

Overall, the provision of ecosystem services by suburban natural areas contributes to the resilience of the city and to the well-being of its inhabitants and visitors. The primary beneficiaries of Bordeaux's ecosystem services are the city's citizens. They can enjoy access to natural areas at reasonable distance from their homes, as well as fauna observation opportunities in patches of biodiversity-rich habitats that are still well preserved.

In addition to their global benefits, carbon storage services are valued by Bordeaux's authorities as they contribute to achieving climate objectives. Both the water company in charge of sanitation and drinking water services and the city's water department benefit from water regulation services as ecosystems contribute to water quality and the achievement of rivers' and surface waters' Good Ecological Status objectives that fall under the European Union Water Framework Directive. Water regulation services also reduce flooding that put citizens and the city's infrastructures at risk.

Ecosystem Service Suppliers

Suppliers of ecosystem services in Bordeaux comprise a wide range of actors who help limit the increasing degradation of ecosystem services due to the city's development. The large Landes forest areas that contribute to carbon storage are sustainably managed both by the public National Office of Forests and by private landowners. Regarding recreation, the Nature Department of Bordeaux Metropolis develops projects that increase the accessibility of public parks and suburban natural areas, through improving the network of walkable paths. These efforts to supply more recreation services are coupled with efforts to enhance biodiversity conservation by protecting aquatic and terrestrial ecological corridors. Local conservation NGOs play a key role in ensuring the viability of these corridors. For instance, one NGO (SEPANSO) manages a small wetland biodiversity-rich natural reserve in the north of Bordeaux, while another (Cistude Nature) owns and protects an old watermill area on a wild part of the river with European pond turtle habitat. The local water company also implemented more biodiversity-friendly catchment management practices in collaboration with local NGOs in several sites under their supervision.

The city, along with the Chamber of Agriculture, also provides incentives to local farmers in order to support traditional small-scale agriculture within the metropolis boundaries to help slow the sale of land for housing development. Maintenance of small-scale agricultural activities and ecological corridors also contributes to water regulation services. Plans for halting the development of large wetland areas were also explored by the city's authorities under the "55,000 ha for nature" initiative.

Terms of the Exchange: Quid Pro Quo

Efforts for ecosystem protections are provided by a diversity of stakeholders, and the metropolis of Bordeaux has developed policies and programs to better take into account nature in urban development. However, there is currently no overall strategy for ecosystem services protection based on measurable results.

The study allowed various actors to better understand the potential of ecosystem services assessments and has been stimulating a common basis of discussion among stakeholders who do not share the same relationship with urban natural areas. The study has remained essentially exploratory in nature, however, and so far has not changed incentives for the protection and restoration of ecosystems within the area.

Mechanism for Transfer of Value

In such urban contexts where multiple layers of land control and a diversity of private and public initiatives coexist, the ability to identify who contributes to the protection of ecosystem services—and by which mechanisms—is fundamental, yet challenging. The study mapped a diversity of projects and activities, planned or already underway by the city's authorities (Climate Plan; 55,000 ha for nature initiative; development of a biodiversity atlas, etc.) and other land managers, which are designed to protect or improve ecosystem services: for example, habitat restoration by NGOs and rivers trusts; environment-friendly agricultural practices by small-scale farmers; biodiversity and soil quality monitoring by research centers (see fig. 17.4).

Effectiveness

This study was the first to assess and map past and future trends for multiple ecosystem services in Bordeaux (fig. 17.5). The maps and graphs of biophysical results were effective in opening a dialogue and exchange of views between stakeholders. This dialogue was also key to interpreting results. The exploration of future scenarios offered a chance to compare the impacts of different planning options on important ecosystem services.

The effectiveness of the study was limited in that it remained exploratory and, as of 2018, has not led to concrete decisions or new commitments that could contribute to improving ecosystem services protection. Although the Urban Planning Department showed interest in this perspective, the ecosystem services maps and indicators produced have not been used in the design of the new local urban plan (http://fichiers.bordeaux-metropole.fr/plu/PLU31_interactif/plu31.html), despite the demonstration that the urban plan scenario is in fact the worst in terms of ecosystem services protection.

The study helped identify the collective action challenges that need to be

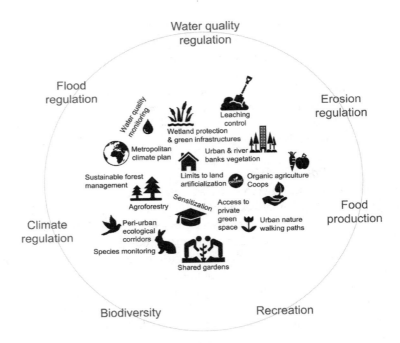

Figure 17.4. Diagram of current projects in Bordeaux that contribute to maintaining or restoring key ecosystem services.

overcome to successfully address ecosystem services degradation. For instance, the powers of Bordeaux's Department of Nature are limited in promoting ecosystem protection as it has to compete with other development priorities promoted by other departments and public institutions (economic and industrial development, housing policies, etc.). Bordeaux's authorities also have to negotiate urban planning choices with the elected representatives of the twenty-eight municipalities who do not necessarily share their interest in ecosystem services protection.

Monitoring and Verification

While no regular monitoring system of Bordeaux's ecosystem services currently exists, this study could serve as a first basis for the development of a more systematic monitoring or accounting system for the city. Such a framework could be used for systematic long-term monitoring of impacts of the

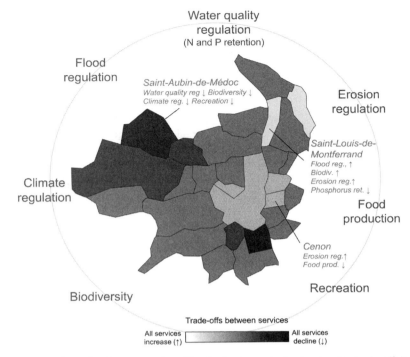

Figure 17.5. Example of a map produced for the Bordeaux study, showing the trade-offs between ecosystem services across different neighborhoods. Adapted from Cabral et al. (2016).

city's development on nature, and on the contributions provided by various stakeholders to its protection.

Key Lessons Learned
The experience in Bordeaux provided key insights for urban planning:

- Ecosystem services assessments at the urban and suburban scale are useful to initiate a rich multi-stakeholders dialogue on the role of ecosystems, and to identify priority areas for ecosystem services protection. Such participatory approaches greatly enrich the development and interpretation of the biophysical results.
- However, at the scale of Bordeaux Metropolis or in smaller municipalities, participatory processes and dialogue did not guarantee that ecosys-

tem services values were formally taken into account in the design and implementation of urban plans.

- It is necessary to bridge the gap between exploratory approaches such as the one used in this study and the concrete negotiations of new responsibilities, commitments, and actions among the multiple stakeholders for ecosystems services protection.
- Addressing this implementation gap requires that alongside ecosystems services assessments, a rigorous social and political context analysis be conducted to identify current projects that already contribute to ecosystem services protection, analyze the priorities and power dynamics at play, and put the ecosystems services results at the service of an effective urban environmental strategy.

Case 3. Urban Agriculture on the Wild Side: Welcome to La Fournillière

The Problem

La Fournillière is a district of the French city of Nantes (fig. 17.3). Historically, the place was occupied by fruit and vegetable canneries that provided kitchen gardens to their employees. During the postwar boom, public housing replaced the canneries and various development projects were proposed. The gardeners were evicted but the projects were eventually abandoned and the garden areas turned into a wasteland of more than three hectares. From the mid-1970s, people living in the neighborhood started to move back to La Fournillière, reestablished some kitchen gardens, and soon enough an informal community was created. Two categories of squatting gardeners worked a plot at La Fournillière: former evicted gardeners, or their children, who were poor but not marginalized people; and residents from the low-income public housing, with very different origins and ethnic backgrounds, many of whom were unemployed and living on social benefits. For them, "owning" a piece of land at La Fournillière was a way of keeping active. It was also a place where they could grow roots, literally. While both groups tended to ignore each other, an element of solidarity brought them together: they were all fully aware of how precarious and uncertain their future was.

In the early 1990s, something unusual happened when the newly elected city council took interest again in La Fournillière with a new neighborhood park project. Rather than requesting to stay and organizing protests, the gar-

Figure 17.6. A kitchen garden in La Fournillière, Nantes. (Photo credit: P. Hamel.)

deners united forces and proposed an alternative project that maintained their gardens. They compiled an in-depth report on La Fournillière, providing exact descriptions and maps of the different pieces of land, spatial pattern of the different gardens, and their history. The report brought to light all the clearing and planting they had done and the public goods they had created. It illustrated the social and ecological value of these gardens for the whole city. The planners of the city of Nantes understood that the opposition of the gardeners was not a negative NIMBY ("not in my backyard") reaction but the expression of collective will and skills. Realizing that an alternative proposal with strong local community support was emerging, they agreed to discuss it with the gardeners' collective. At the end of a long process of negotiations, the city council decided to support the gardeners' alternative project in place of the original proposal.

La Fournillière is now a park organized around islets or patches formed by the kitchen gardens that previously existed on the site (fig. 17.6). Paths for walkers and runners intertwine with the islets and connect them. At the center, a visitor and education center initiates visitors to the recycling of material and waste in urban gardening, including waste sorting and composting to enhance biomass and biodiversity. A maze of service alleys spreads around five key items: three wells, a pond, and an improvised "pétanque" (French lawn bowl sport) court.

The Ecosystem Services

The site of La Fournillière provides many ecosystem services: it produces local food, enriches urban biodiversity, and helps maintain pleasant air temperatures. Beyond the biophysical benefits, La Fournillière is also a leisure area where people can stroll or run. It is also a major asset to strengthen the social fabric and empower local communities.

Ecosystem Service Beneficiaries

The services of La Fournillière benefit the gardeners, other citizens, and the local authorities. First, the gardeners value their kitchen gardens as a place where they can retreat and be creative in how they decorate the gardens with things that would not be permitted anywhere else in the city (for example, a doll's head impaled on a pole, a teddy bear crucified on a picket fence, etc.). The kitchen gardens also have direct economic benefits: cabbages, potatoes, and other vegetables are planted and feed families year-round. Second, other citizens also benefit from the park as a recreation area and a cooling island during heat waves. Enjoying a prime location in the city, La Fournillière is also a shortcut for many people going to school, to work, or to the market in Nantes. Finally, local authorities enjoy several benefits from the site: the kitchen gardens help reduce social tensions within the nearby public housing area, and La Fournillière is an initiative they can showcase that combines both bottom-up and top-down processes of decision making in the urbanization process. The park also gives consistency to the urban fabric by connecting neighborhoods together.

Ecosystem Service Suppliers

The main suppliers are the gardeners, who created this place and now dictate the rules within the park: more efficient and wiser water management; a ban on cutting any tree in one's own gardens, since trees are considered to be shared assets, and so on.

The second supplier is the municipality of Nantes, which decided to enact the gardeners' initiative and legalize it, recommoning urban land through agriculture.

Terms of the Exchange: Quid Pro Quo

Legally La Fournillière belongs to the city of Nantes: it is considered a municipal park, which means that the city pays for its maintenance. The gardeners do not pay for their plot, but in exchange for their plot they commit themselves to create and run a museum to educate visitors about the recycling of material and waste in urban gardening. The gardeners also commit to help the municipality to maintain the park, and to ensure open access to the many lanes that

run across the lots. Gardeners are organized under a not-for-profit coopera-
tive, which manages the new gardeners and provides them with plots.

Monitoring and Verification
To our knowledge, there is no formal monitoring or assessment of the site's
management conducted by the gardeners' association or the city.

Effectiveness
The direct health and well-being benefits of the site, from food production to
recreation, have not been quantified but are likely to be substantial based on
the assessment of similar urban parks. Urban agriculture is a common good,
bringing people together and providing a pathway toward reintegrating nature in cities. These benefits are difficult to measure, but they contribute to
environmental stewardship and positive relationships between citizens and the
planning authorities.

Key Lessons Learned
The example of La Fournillière shows how a squatted wasteland in a poor place
near social housing blocks can be turned into a very popular park based on urban agriculture. It embodies a bottom-up approach to urban planning and also
symbolizes the potential of participatory approaches to planning procedures,
that is, bringing everyone to the table so that citizens understand that urban
affairs are fundamentally their affairs.

Urban agriculture lends itself particularly well to long-lasting urban policies, especially those turning brownfields and wastelands into environmental
goods and urban amenities. Urban agriculture in interstitial abandoned urban
areas can improve resilience and provide multiple direct benefits. The example
of La Fournillière shows that seeds of urban agriculture may be present in the
urban fabric and that participatory approaches to urban planning can transform them into full-grown solutions to promote urban sustainability.

Conclusions
The case studies show that nature-based solutions can provide multiple services, from efficient water management to food production, climate mitigation,
recreation opportunities, and biodiversity protection. They suggest that the
multiscale, multibenefit nature of nature-based solutions is both an advantage

to addressing multiple challenges in complex systems, and a potential barrier to its integration in existing planning processes. In the following sections, we discuss these findings by summarizing the types of mechanisms that support the implementation of nature-based solutions in cities. We conclude with the barriers to implementation and management options to overcome those.

Mechanisms to Incorporate Nature into Urban Planning

Promoting nature-based solutions often requires collaboration between city departments (engineering, planning, housing, etc.) to implement formal "top-down" mechanisms, together with participatory approaches and the integration of "bottom-up" initiatives into formal planning.

Top-down Mechanisms

Through regulations and economic incentives, cities can directly influence the uptake of nature-based solutions. As demonstrated in the Yarra Range Council example, the infrastructure department, in collaboration with water agencies, provides guidance and enforces regulations that promote nature-based stormwater management techniques. Another example is building codes, through which the city can impose some minimum levels of adoption of nature-based solutions. In Melbourne, such a strategy was adopted under the 2012 planning amendment, where new buildings had to comply with ambitious energy- or water-efficiency requirements.

In parallel, cities can adopt soft measures to raising awareness, building capacity for professionals, and encouraging private sector investments. These measures can take the form of strategic plans (e.g., Yarra Ranges' integrated urban water management plan, Paris' rainwater plan), professional forums, and multiple forms of incentives for companies to invest and innovate in nature-based solutions. In the San Francisco Bay Area, for example, an ambitious *Resilient by Design* competition was launched to encourage professionals and the civil society to join forces and address the climate challenges in the region.

Bottom-up Engagement

A city does not arise from the sole will and skill of architects, planners, and politicians. It is also molded by its citizens. Such a process takes time, often operating quite differently from timelines of elected officials and planners who are constrained by factors such as election cycles and construction deadlines. In this context, nature-based solutions lend themselves to the promotion of participatory approaches and urban "tinkering," that is, the recognition that unplanned, citizen-led initiatives form part of urban life and can effectively contribute to urban sustainability.

In our case studies, the example of La Fournillière exemplified this slow, grass-roots process, whereby urban agriculture was first initiated by illegal squatters, and later adopted in the city's plans in recognition of the social and ecological capital built by the community. In the San Francisco Bay Area (see case 3, chapter 14), multiple organizations are working to promote nature-based solutions, from environmental NGOs to research institutes studying the multiple benefits of adaptation strategies. The participation of research and academic communities was also a key element of the Yarra Ranges and Bordeaux case studies. In both cases, it helped build trust and lend credibility to the project. In Yarra Ranges, it is also an integral part of the strategy to continuously monitor and improve practices to achieve better results.

Navigating the Complexity of the Social, Ecological, and Technical Interactions

The case studies demonstrate the multiple benefits of nature-based solutions, from both biophysical and social perspectives. The case of stormwater in Melbourne illustrates the direct environmental and economic benefits that nature-based solutions can provide. The Bordeaux case study illustrates that broadening the performance metrics to include "invisible" services provided by nature may change the discourse on regional planning. In other cases, like urban agriculture in Nantes, the main benefits beyond the production of food—relating to the protection of social capital—may be the most difficult to measure.

While our examples were selected to illustrate important benefits of nature-based solutions, they do not imply, of course, that nature-based solutions should be implemented at all costs and in any situation. Social, technical, and ecological factors determine the performance of such solutions compared to more conventional alternatives. Local actors need to join forces to determine, in a given place and time, which solutions are adequate. The various policy or social mechanisms described above can then be used to implement the most relevant solutions, with a mix of bottom-up and top-down approaches.

From a governance standpoint, approaches to build socioecological resilience and promote adaptive planning offer practical recommendations to navigate the complexity of the concept of nature in the city. The general approach is to promote complex-systems thinking and revisit strategies regularly as social, technical, and ecological factors change. Although this departs from traditional planning approaches, cities seem well positioned to drive these social transformations and a number of local government initiatives, such as C40 (www.c40 .org); ICLEI (www.iclei.org); and 100 Resilient Cities (www.100resilientcities .org), already prove their leadership for a more sustainable world.

Key References

d'Amour, Christopher Bren, Femke Reitsma, Giovanni Baiocchi, Stephan Barthel, Burak Güneralp, Karl-Heinz Erb, Helmut Haberl, Felix Creutzig, and Karen C. Seto. 2017. "Future urban land expansion and implications for global croplands." *Proceedings of the National Academy of Sciences* 114, no. 34:8939–44.

Cabral, Pedro, Clément Feger, Harold Levrel, Mélodie Chambolle, and Damien Basque. 2016. "Assessing the impact of land-cover changes on ecosystem services: A first step toward integrative planning in Bordeaux, France." *Ecosystem Services* 22:318–27.

Feger, Clément, and Laurent Mermet. 2017. "A blueprint towards accounting for the management of ecosystems." *Accounting, Auditing & Accountability Journal* 30, no. 7:1511–36.

Keeler, B. L., et al. Forthcoming. Context-dependency of urban ecosystem service values. *Nature Sustainability.*

Lefebvre, Henri. 1968. *Le Droit à la Ville.* Paris: Anthropos.

Levrel, Harold, Pedro Cabral, Clément Feger, Mélodie Chambolle, and Damien Basque. 2017. "How to overcome the implementation gap in ecosystem services? A user-friendly and inclusive tool for improved urban management." *Land Use Policy* 68:574–84.

Magnaghi, A. 2005. *The Urban Village: A Charter for Democracy and Local Self-Sustainable Development.* London: Zed Books.

Mancebo, François. 2018. "Gardening the city: Addressing sustainability and adapting to global warming through urban agriculture." *Environments* 5, no. 3:38.

Pasquier, E. 2001. *Cultiver son jardin: Chroniques des jardins de la Fournillière, 1992–2000.* Paris: L'Harmattan.

Seto, Karen C., Burak Güneralp, and Lucy R. Hutyra. 2012. "Global forecasts of urban expansion to 2030 and direct impacts on biodiversity and carbon pools." *Proceedings of the National Academy of Sciences* 109, no. 40:16083–88.

Viljoen, A. and J. Howe. 2005. *Continuous Productive Urban Landscapes.* London: Routledge.

Yarra Ranges Council. 2017. *Integrated Water Management Plan 2017.* www.yarraranges.vic .gov.au.

Acknowledgments

We are deeply indebted to everyone involved in the efforts we capture here. Too numerous to name individually, these are the heroes transforming ideas into action and making the world a better place for both people and nature. Our hope is that this book serves to amplify their stories and efforts. Included in this group are the many people and institutions involved in the Natural Capital Project, founded in 2005. It is through this global partnership that many of the experiences we present emerged.

We are extremely grateful to Erin Johnson and Island Press for support in creating this book, and for encouragement, patience, and wise counsel throughout the writing, editing, and production process. We are also grateful for the generous contributions of the many people who shared their expertise and provided feedback on particular chapters and case studies. They include Nicola Virgill-Rolle from the Office of the Prime Minister, The Bahamas; Marc Diaz from NatureVest at The Nature Conservancy (TNC); Emily McKenzie from World Wildlife Fund (WWF); Fredrick Kihara and Colin Apse from TNC Africa program; David Le Maitre and Nadia Sitas from the Council for Scientific and Industrial Research, South Africa; Rob Pacheco of Hawaii Forest & Trail; Sam Ham from University of Idaho; and Gregory Verutes, previously from the Natural Capital Project. We also thank Christa Anderson from Stanford University; Michele Dailey from WWF–US; Emily Chapin from TNC; Katie Chang from Land Trust Alliance; and WWF Brazil, for generously sharing data for maps and figures. From the Natural Capital Project, Sarah Cafasso assisted with editing, and Charlotte Weil designed many of the maps and figures.

None of this would have been possible without the many funders who have supported our vision and efforts over many years. The Paulson Institute funded the early concept of this book (published in Chinese). We are especially grateful to Peter and Helen Bing, and Vicki and Roger Sant, who believed in and encouraged us and the growth of this movement from the very beginning.

Contributors

Stuart Anstee, Stuart Anstee and Associates, Australia

Katie Arkema, Natural Capital Project, Stanford University and University of Washington, United States

Ian Bateman, Land, Environment, Economics and Policy Institute, University of Exeter Business School, United Kingdom

Onon Bayasgalan, Wildlife Conservation Society, Mongolia

Silvia Benitez, The Nature Conservancy, Latin America Region

Rebecca Benner, The Nature Conservancy, New York State, United States

Rick Bennett, Northeast Region, US Fish and Wildlife Service, United States

Nirmal Bhagabati, World Wildlife Fund, United States

Amy Binner, Land, Environment, Economics and Policy Institute, University of Exeter Business School, United Kingdom

Carter Brandon, World Bank

Kate A. Brauman, Institute on the Environment, University of Minnesota, United States

Leah Bremer, University of Hawaiʻi Economics Research Organization and Water Resources Research Center, United States

Benjamin P. Bryant, Water in the West and the Natural Capital Project, Woods Institute for the Environment, Stanford University, United States

Rebecca Chaplin-Kramer, Natural Capital Project, Woods Institute for the Environment, Stanford University, United States

Beatrice E. Crona, Global Economic Development and the Biosphere Academy Program (GEDB); Royal Swedish Academy of Sciences; and Stockholm Resilience Centre, Stockholm University, Sweden

Helen Crowley, Sustainable Sourcing Innovation, Kering, France

Gretchen C. Daily, Natural Capital Project, Department of Biology and Stanford Woods Institute, Stanford University, United States

Alyssa Dausman, The Water Institute of the Gulf, United States

Brett Day, Land, Environment, Economics and Policy Institute, University of Exeter Business School, United Kingdom

Clément Feger, Montpellier Research in Management, University of Montpellier; and AgroParisTech, France

Marcus Feldman, Department of Biology, Stanford University, United States

Lauren Ferstandig, NatureVest, The Nature Conservancy, United Kingdom

Carlo Fezzi, Land, Environment, Economics and Policy Institute, University of Exeter Business School, United Kingdom

Carl Folke, Beijer Institute of Ecological Economics and Global Economic Development and the Biosphere Academy Program, Royal Swedish Academy of Sciences; and Stockholm Resilience Centre, Stockholm University, Sweden

Victor Galaz, Beijer Institute of Ecological Economics, Royal Swedish Academy of Sciences; and Stockholm Resilience Centre, Stockholm University, Sweden

Line J. Gordon, Stockholm Resilience Centre, Stockholm University, Sweden

Anne D. Guerry, Natural Capital Project, Woods Institute for the Environment, Stanford University, United States

Perrine Hamel, Natural Capital Project, Woods Institute for the Environment, Stanford University, United States

Stéphanie Hamel, Yarra Ranges Council, VIC, Australia

Hanna Helsingen, WWF, Myanmar

Craig Holland, The Nature Conservancy, United States

Annette Killmer, Inter-American Development Bank

Zachary Knight, Blue Forest Conservation

Linqiao Kong, Research Center for Eco-Environmental Sciences, Chinese Academy of Sciences, China

Eric F. Lambin, School of Earth, Energy and Environment Sciences and Woods Institute for the Environment, Stanford University, United States; and Georges Lemaître Earth and Climate Research Centre, Earth and Life Institute, University of Louvain, Belgium

Jim Leape, Center for Ocean Solutions, Woods Institute for the Environment, Stanford University, United States

Kai Lee, Center for Ocean Solutions, Woods Institute for the Environment, Stanford University, United States

Michele Lemay, Inter-American Development Bank (retired)

Cong Li, Institute of Population and Development Studies, Xi'an Jiaotong University, China

Jie Li, Institute of Population and Development Studies, Xi'an Jiaotong University, China

Shuzhuo Li, Institute of Population and Development Studies, Xi'an Jiaotong University, China

Jianguo Liu, Center for Systems Integration and Sustainability, Department of Fisheries and Wildlife, Michigan State University, United States

François Mancebo, Centre of Research on Law and Territory, IATEUR (Institute of Regional Development and Sustainable Urban Planning); University of Reims and IRCS (International Research Center on Sustainability) of Reims, France

Lisa Mandle, Natural Capital Project, Woods Institute for the Environment, Stanford University, United States

Len Materman, San Francisquito Creek Joint Powers Authority, United States

Sai Nay Won Myint, WWF, Myanmar

Carl Obst, Institute for Development of Environmental-Economic Accounting; and the University of Melbourne, Australia

Henrik Österblom, Stockholm Resilience Centre, Stockholm University, Sweden

Zhiyun Ouyang, Research Center for Eco-Environmental Sciences, Chinese Academy of Sciences, China

Stephen Polasky, Natural Capital Project, Institute on the Environment; Department of Ecology, Evolution and Behavior, and Department of Applied Economics, University of Minnesota, United States

Mary Ruckelshaus, Natural Capital Project, Woods Institute for the Environment, Stanford University, United States

Alex Rusby, Harrow School, United Kingdom

Phil Saksa, Blue Forest Conservation

James Salzman, Bren School of Environmental Science and Management, University of California, Santa Barbara; and University of California Los Angeles School of Law, United States

Lisen Schultz, Stockholm Resilience Centre, Stockholm University, Sweden

Enkhtuvshin Shiilegdamba, Wildlife Conservation Society, Mongolia

Greg Smith, Land, Environment, Economics and Policy Institute, University of Exeter Business School, United Kingdom

Changsu Song, Research Center for Eco-Environmental Sciences, Chinese Academy of Sciences, China

Meg Symington, World Wildlife Fund, United States

Rick Thomas, Bren School of Environmental Science and Management, University of California, Santa Barbara, United States

Samdanjigmed Tulganyam, Oyu Tolgoi LLC

Alvaro Umaña Quesada, Centro Agronómico Tropical de Investigación y Enseñanza (CATIE), Costa Rica

Ray Victurine, Wildlife Conservation Society

Kari Vigerstøl, The Nature Conservancy, Global Water

Bhaskar Vira, Department of Geography and Conservation Research Institute, University of Cambridge, United Kingdom

Greg Watson, Inter-American Development Bank

Charlotte Weil, Natural Capital Project, Woods Institute for the Environment, Stanford University, United States

Ruth Welters, School of Environmental Sciences, University of East Anglia, United Kingdom

Nick Wobbrock, Blue Forest Conservation

Christina Wong, Research Center for Eco-Environmental Sciences, Chinese Academy of Sciences, China

Weihua Xu, Research Center for Eco-Environmental Sciences, Chinese Academy of Sciences, China

Hua Zheng, Research Center for Eco-Environmental Sciences, Chinese Academy of Sciences, China

About the Editors

Lisa Mandle is a lead scientist at the Natural Capital Project at Stanford University. Her research is focused on the impacts of land management and infrastructure development on ecosystem service provision and on the social and economic equity dimensions of ecosystem service benefits. She is also leading new science in ecosystem change and human health, with practical applications.

Dr. Mandle has worked with governments, multilateral development banks, and nongovernmental organizations to incorporate this understanding into development decisions, particularly in Latin America and Asia. She led development of guidance for the Inter-American Development Bank on integrating natural capital into road planning and investment, and of a decision-support software tool for biodiversity and ecosystem service offsets in Colombia. She has also led trainings around the world on natural capital-based approaches and tools for decision making.

Zhiyun Ouyang is professor and director of Research Center for Eco-Environmental Sciences, Chinese Academy of Sciences. He leads unparalleled efforts to bring understanding of the earth system to bear on crucial societal issues of extreme poverty and environmental degradation research in a wide range of areas. He is focused principally on ecosystem services, ecosystem assessment, and ecological planning from city to national scales; ecosystem restoration and biodiversity conservation, including the design of China's new national park system; and policy and finance mechanisms for achieving inclusive green growth. He has opened the science of ecosystem services, ecosystem restoration, and biodiversity conservation and moved it into real-world practice at scale, in China and beyond.

Dr. Ouyang has published over a dozen books and hundreds of scientific articles and reports, including more than 150 papers in international journals. His books include *Developing Gross Ecosystem Product (GEP) and Ecological Asset Accounting for Eco-compensation* (2017); *Ecological Security Strategy* (2014); and *Regional Ecosystem Assessment and Ecosystem Service Zoning*

(2009). He serves as the president of the Ecological Society of China and as a board member of the International Association of Ecology. He has won three national Science and Technology Achievement awards in China.

James Salzman is the Donald Bren Distinguished Professor of Environmental Law with joint appointments at the UCLA School of Law and the Bren School of Environmental Science and Management, UC Santa Barbara. He was the first scholar to address the legal and institutional aspects of creating markets for ecosystem services. He has worked with governments in Australia, Canada, China, India, New Zealand, and other countries to help design their payments for ecosystem services programs.

Dr. Salzman's research ranges from drinking water and policy instrument design to conservation and trade conflicts. Author of over ninety articles and ten books, his publications have been downloaded over 100,000 times. He is active on environmental boards and in government policy bodies and serves on both the National Drinking Water Advisory Council (reporting to the US Environmental Protection Agency) and the Trade and Environment Policy Advisory Committee. His previous book, *Drinking Water: A History* (2012), is very widely read.

Gretchen C. Daily is Bing Professor of Environmental Science at Stanford University, where she also serves as Senior Fellow in the Stanford Woods Institute for the Environment; director of the Center for Conservation Biology; and cofounder and faculty director of the Natural Capital Project, a global partnership driving innovation to value nature explicitly and systematically in policy, finance, and management. Dr. Daily's interdisciplinary research is focused on harmonizing people and nature in biodiversity dynamics and conservation, land use and agriculture, and human livelihoods; the production and value of ecosystem support for human health, prosperity, and overall well-being; and policy and finance innovation for achieving inclusive green growth. She works with decision makers in key contexts worldwide, codeveloping practical tools and widely shared approaches.

Dr. Daily has published hundreds of scientific and popular articles. Her dozen books include *Nature's Services: Societal Dependence on Natural Ecosystems* (Island Press 1997); *The New Economy of Nature: The Quest to Make Conservation Profitable* (Island Press 2002); *Natural Capital: Theory and*

Practice of Mapping Ecosystem Services (2011); *Research on Rural Household Livelihood and Environmental Sustainable Development* (2017, in Chinese), and *One Tree* (2018). She is a fellow of the US National Academy of Sciences, the American Academy of Arts and Sciences, and the American Philosophical Society.

Index

Island Press | Board of Directors

Pamela Murphy
(Chair)

Terry Gamble Boyer
(Vice Chair)
Author

Tony Everett
(Treasurer)
Founder,
Hamill, Thursam & Everett

Deborah Wiley
(Secretary)
Chair, Wiley Foundation, Inc.

Decker Anstrom
Board of Directors,
Discovery Communications

Melissa Shackleton Dann
Managing Director,
Endurance Consulting

Margot Ernst

Alison Greenberg

Rob Griffen
Managing Director,
Hillbrook Capital

Marsha Maytum
Principal,
Leddy Maytum Stacy Architects

David Miller
President, Island Press

Georgia Nassikas
Artist

Alison Sant
Cofounder and Partner,
Studio for Urban Projects

Ron Sims
Former Deputy Secretary,
US Department of Housing
and Urban Development

Sandra E. Taylor
CEO, Sustainable Business
International LLC

Anthony A. Williams
CEO and Executive Director,
Federal City Council

Sally Yozell
Senior Fellow and Director
of Environmental Security,
Stimson Center